This report contains the collective views of an international group of experts and does not necessarily represent the decisions or the stated policy of the United Nations Environment Programme, the International Labour Organisation, or the World Health Organization

Environmental Health Criteria 64

CARBAMATE PESTICIDES: A GENERAL INTRODUCTION

Published under the joint sponsorship of
the United Nations Environment Programme,
the International Labour Organisation,
and the World Health Organization

World Health Organization
Geneva, 1986

The International Programme on Chemical Safety (IPCS) is a joint venture of the United Nations Environment Programme, the International Labour Organisation, and the World Health Organization. The main objective of the IPCS is to carry out and disseminate evaluations of the effects of chemicals on human health and the quality of the environment. Supporting activities include the development of epidemiological, experimental laboratory, and risk-assessment methods that could produce internationally comparable results, and the development of manpower in the field of toxicology. Other activities carried out by IPCS include the development of know-how for coping with chemical accidents, coordination of laboratory testing and epidemiological studies, and promotion of research on the mechanisms of the biological action of chemicals.

ISBN 92 4 154264 0

The designations employed and the presentation of the material in this publication do not imply the expression of any opinion whatsoever on the part of the Secretariat of the World Health Organization concerning the legal status of any country, territory, city or area or of its authorities, or concerning the delimitation of its frontiers or boundaries.

The mention of specific companies or of certain manufacturers' products does not imply that they are endorsed or recommended by the World Health Organization in preference to others of a similar nature that are not mentioned. Errors and omissions excepted, the names of proprietary products are distinguished by initial capital letters.

ISSN 0250-863X

PRINTED IN FINLAND

DHSS — VAMMALA — 6000

CONTENTS

Page

ENVIRONMENTAL HEALTH CRITERIA FOR CARBAMATE PESTICIDES: A GENERAL INTRODUCTION

1. SUMMARY AND RECOMMENDATIONS 11

1.1 Summary. 11
1.1.1 General 11
1.1.2 Properties, uses, and analytical methods 12
1.1.3 Sources, environmental transport and distribution 12
1.1.4 Environmental levels and exposures . . . 12
1.1.5 Effects on organisms in the environment . 13
1.1.6 Kinetics and metabolism 13
1.1.7 Mechanism of toxicity 14
1.1.8 Effects on experimental animals and *in vitro* test systems 14
1.1.9 Mutagenicity and related end-points . . . 16
1.1.10 Carcinogenicity 16
1.1.11 Effects on man 17
1.1.12 Previous evaluations by international bodies 17
1.2 Recommendations 17

2. IDENTITY, PHYSICAL AND CHEMICAL PROPERTIES, ANALYTICAL METHODS 20

2.1 Identity 20
2.2 Physical and chemical properties 20
2.3 Analytical methods 21

3. SOURCES OF HUMAN AND ENVIRONMENTAL EXPOSURE 26

3.1 Natural occurrence 26
3.2 Man-made sources 26
3.2.1 Production levels, processes, and uses . 26

4. ENVIRONMENTAL TRANSPORT, DISTRIBUTION, AND TRANSFORMATION 28

4.1 Transport and distribution between media 28
4.1.1 Air 28
4.1.2 Water 28
4.1.3 Soil 28
4.1.4 Vegetation and wildlife 30

Page

4.1.5 Entry into the food chain 31
4.2 Biotransformation 31
4.2.1 Microbial degradation 31
4.2.2 Photodegradation 32
4.2.3 Photodecomposition in the aquatic environment 32

5. ENVIRONMENTAL LEVELS AND HUMAN EXPOSURE 34

5.1 Exposure of the general population 34
5.1.1 Food and drinking-water 34

6. METABOLISM AND MODE OF ACTION 35

6.1 Mode of action 35
6.1.1 Neuropathy target esterase 37
6.2 Metabolism 38
6.2.1 Absorption 38
6.2.2 Distribution 38
6.2.3 Metabolic transformation 39
6.2.3.1 Biotransformation mechanisms . . 39
6.2.3.2 Oxidation 39
6.2.3.3 Hydrolysis 41
6.3.2.4 Conjugation 41
6.2.3.5 Examples of biotransformation of carbamates 41
6.3 Elimination and excretion in expired air, faeces, and urine 45
6.3.1 Man 45
6.3.2 Laboratory animals 46
6.4 Metabolism in plants 47

7. EFFECTS ON ORGANISMS IN THE ENVIRONMENT 49

7.1 Microorganisms 49
7.2 Aquatic Organisms 49
7.2.1 Field studies 54
7.3 Terrestrial Organisms 55
7.3.1 Effects on soil fauna 55
7.3.2 Wildlife 55
7.3.3 Bees 56
7.4 Earthworm and mite populations 56

8. EFFECTS ON EXPERIMENTAL ANIMALS AND _IN VITRO_ TEST SYSTEMS 58

8.1 Single exposures 58

Page

8.1.1 Oral 58
8.1.2 Dermal 62
8.1.3 Inhalation 62
8.2 Short- and long-term exposures 62
8.2.1 Oral 63
8.2.1.1 Further information on short- and long-term toxicity 65
8.2.2 Dermal 66
8.3 Skin and eye irritation; sensitization 66
8.3.1 Skin irritation 66
8.3.2 Eye irritation 66
8.3.3 Skin sensitization 67
8.4 Inhalation 67
8.5 Reproduction, embryotoxicity, and teratogenicity 68
8.5.1 Reproduction 68
8.5.2 Endocrine system 70
8.5.3 Embryotoxicity and teratogenicity 72
8.6 Mutagenicity and related end-points 75
8.7 Carcinogenicity 80
8.7.1 General 80
8.8 Special studies 83

9. EFFECTS ON MAN 84

9.1 General population exposure 84
9.1.1 Acute toxicity: poisoning incidents . . . 84
9.1.2 Effects of short- and long-term exposure 84
9.1.3 Controlled human studies 85
9.2 Occupational exposure 86
9.2.1 Acute toxicity: poisoning incidents . . . 86
9.2.2 Effects of short- and long-term exposure 88
9.2.3 Epidemiological studies 89
9.3 Signs and symptoms of acute intoxication by carbamates 89
9.3.1 Biochemical methods for measurement of effects 90
9.4 Treatment of acute poisoning by carbamate insecticides 91
9.4.1 Minimizing the absorption 91
9.4.2 General supportive treatment 92
9.4.3 Specific pharmacological treatment . . . 92
9.4.3.1 Atropine 92
9.4.3.2 Oxime reactivators 92
9.4.3.3 Diazepam 92

		Page
10. PREVIOUS EVALUATIONS BY INTERNATIONAL BODIES		93
REFERENCES		94
ANNEX I	NAMES AND STRUCTURES AND SOME PHYSICAL AND CHEMICAL PROPERTIES OF CARBAMATE PESTICIDES .	119
ANNEX II	SUMMARY OF THE SHORT- AND LONG-TERM TOXICITY STUDIES THAT WERE USED TO ESTABLISH THE ACCEPTABLE DAILY INTAKES FOR HUMAN BEINGS FOR CARBAMATE COMPOUNDS	128
ANNEX III	CARBAMATES: JMPR REVIEWS, ACCEPTABLE DAILY INTAKES, EVALUATION BY IARC, CLASSIFICATION BY HAZARD, FAO/WHO DATA SHEETS, IRPTC DATA PROFILE AND LEGAL FILE	130
ANNEX IV	ABBREVIATIONS	136

WHO TASK GROUP ON CARBAMATE PESTICIDES

Members

Dr D. Ecobichon, Department of Pharmacology and Therapeutics, McGill University, Montreal, Quebec, Canada

Dr A.H. El-Sebae, Department of Pesticide Chemistry, Faculty of Agriculture, University of Alexandria, Alexandria, Egypt

Dr L. Ivanova-Chemishanska, Institute of Hygiene and Occupational Health, Medical Academy, Sofia, Bulgaria (Vice-Chairman)

Dr M.K. Johnson, Toxicology Unit, Medical Research Council Laboratories, Carshalton, Surrey, United Kingdom

Dr S.K. Kashyap, National Institute of Occupational Health, Ahmedabad, India

Dr M. Lotti, Institute of Occupational Health, Padua, Italy

Dr L. Martson, All-Union Scientific Research Institute of the Hygiene and Toxicology of Pesticides, Polymers, and Plastics (VNIIGINTOX), Kiev, USSR[a]

Dr U.G. Oleru, College of Medicine, University of Lagos, Lagos, Nigeria.

Dr W.O. Phoon, Department of Social Medicine and Public Health, National University of Singapore, Outram Hill, Republic of Singapore (Chairman)

Dr A.F. Rahde, Ministry of Public Health, Porto Alegre, Brazil

Dr E. Reiner, Institute for Medical Research and Occupational Health, Zagreb, Yugoslavia

Dr J. Sekizawa, National Institute of Hygienic Sciences, Tokyo, Japan

Observers

Mr R.J. Lacoste, International Group of National Associations of Pesticide Manufacturers (GIFAP), Brussels, Belgium

---

a Invited, but unable to attend.

Secretariat

Mrs B. Bender, United Nations Environment Programme, International Register of Potentially Toxic Chemicals, Geneva, Switzerland

Dr J.R.P. Cabral, Unit of Mechanisms of Carcinogenesis, International Agency for Research on Cancer, Lyons, France

Dr K.W. Jager, International Programme on Chemical Safety, World Health Organization, Geneva, Switzerland (Secretary)

Dr G.J. Van Esch, Bilthoven, The Netherlands (Temporary Adviser) (Rapporteur)

Dr C. Xintaras, Office of Occupational Health, World Health Organization, Geneva, Switzerland

NOTE TO READERS OF THE CRITERIA DOCUMENTS

Every effort has been made to present information in the criteria documents as accurately as possible without unduly delaying their publication. In the interest of all users of the environmental health criteria documents, readers are kindly requested to communicate any errors that may have occurred to the Manager of the International Programme on Chemical Safety, World Health Organization, Geneva, Switzerland, in order that they may be included in corrigenda, which will appear in subsequent volumes.

* * *

Detailed data profiles and legal files for most of the carbamate pesticides can be obtained from the International Register of Potentially Toxic Chemicals, Palais des Nations, 1211 Geneva 10, Switzerland (Telephone no. 988400 - 985850).

ENVIRONMENTAL HEALTH CRITERIA FOR CARBAMATE PESTICIDES

A WHO Task Group on Environmental Health Criteria for Carbamate Pesticides met in Geneva from 30 September to 4 October 1985. Dr K.W. Jager opened the meeting on behalf of the Director-General. The Task Group reviewed and finalized the draft criteria document.

The efforts of all who helped in the preparation and finalization of the document are gratefully acknowledged.

* * *

Partial financial support for the publication of this criteria document was kindly provided by the United States Department of Health and Human Services, through a contract from the National Institute of Environmental Health Sciences, Research Triangle Park, North Carolina, USA - a WHO Collaborating Centre for Environmental Health Effects. The United Kingdom Department of Health and Social Security generously supported the cost of printing.

1. SUMMARY AND RECOMMENDATIONS

1.1 Summary

1.1.1 General

The carbamates discussed in this publication are those mainly used in agriculture, as insecticides, fungicides, herbicides, nematocides, or sprout inhibitors. In addition, they are used as biocides for industrial or other applications and in household products. A potential use is in public health vector control. Thus, these chemicals are part of the large group of synthetic pesticides that have been developed, produced, and used on a large scale in the last 40 years.

The general formula of the carbamates is:

$$R^1NH - \overset{\overset{\displaystyle O}{\|}}{C} - OR^2$$

where R^1 and R^2 are alkyl or aryl groups.

More than 50 carbamates are known, and it is clear that it is not within the scope of this introduction to include all the information about each compound. However, all the different aspects of the different classes of carbamates are touched on, making use of available publications and reports of studies and using well known carbamates, such as carbaryl, benomyl, and a few others, as examples.

Thiocarbamates and dithiocarbamates have not been included, because these groups of compounds have a different mode of action and will be dealt with in a separate publication.

1.1.2 Properties, uses, and analytical methods

Three classes of carbamate pesticides are known. The carbamate ester derivatives, used as insecticides (and nematocides), are generally stable and have a low vapour pressure and low water solubility. The carbamate herbicides (and sprout inhibitors) have the general structure $R^1NHC(O)OR^2$, in which R^1 and R^2 are aromatic and/or aliphatic moieties. Carbamate fungicides contain a benzimidazole group.

The only physical or chemical properties given in this publication are the relative molecular mass, vapour pressure, and water solubility.

Analytical methods for a number of important carbamates have been tabulated.

1.1.3 Sources, environmental transport and distribution

The synthesis and commercialization of the carbamate pesticides has been in progress since the 1950s. The benzimidazole fungicides were introduced on the market in about 1970.

In general, the vapour pressure of the carbamates is low; nevertheless, they will evaporate or sublimate slowly at normal temperature, which may lead to volatilization of carbamates from water and soil. However, distribution via air will be a minor factor. The aqueous environment will be an important route of transport for highly-soluble carbamates. The light absorption characteristics of carbamates contribute to their rapid decomposition (by photodegradation or photodecomposition) under aqueous conditions. Thus, the hazards of long-term contamination with carbamates seem small. Carbamate insecticides are mainly applied on the plants, and can reach the soil, while carbamate nematocides and herbicides are applied directly to the soil. Several factors influence the biodegradation of carbamates in soil, such as volatility, soil type, soil moisture, adsorption, pH, temperature, and photodecomposition. Because the different carbamates have different properties, it is clear that each should be evaluated on its own merits, and no extrapolation of results can be made from one carbamate to another. One carbamate may be easily decomposed, while another may be strongly adsorbed on soil. Some leach out easily and may reach groundwater. In these processes, the soil type and water solubility are of great importance. Furthermore, it should be recognized that this not only concerns the parent compound but also the breakdown products or metabolites.

Environmental conditions that favour the growth and activity of microorganisms also favour the degradation of carbamates. The first step in the metabolic degradation of carbamates in soil is hydrolysis. The hydrolysis products will be further metabolized in the soil-plant system.

Plants can metabolize carbamates in which arylhydroxylation and conjugation, or hydrolytic breakdown are the main routes of detoxification. The results of a number of studies suggest that carbamates are exclusively distributed via the apoplastic system in plants.

Carbamates are metabolized by microorganisms, plants, and animals or broken down in water and soil.

1.1.4 Environmental levels and exposures

From the available data, it appears that bioaccumulation in the different species and different food chains will only take place to a slight extent. Certain carbamates may reach groundwater and as a consequence may find their way into drinking-water. The few studies available indicate that exposure of the general population is low. However, this should be confirmed by market-basket and/or total-diet studies.

1.1.5 Effects on organisms in the environment

Soil microorganisms are capable of metabolizing (hydrolysing) carbamates and can easily adapt themselves to metabolize the different types of carbamates. Nevertheless, carbamates and their metabolites can, at high dose levels, affect the microflora and cause changes that may be of importance in soil productivity. Although carbamates are not very stable under aquatic conditions, and will not persist long in this environment, some bioaccumulate in fish, mainly because the metabolism is slow in fish. Other carbamates are metabolized rapidly and no accumulation occurs. Some carbamates are highly toxic for invertebrates and fish, others much less so. In certain cases, the use of toxic carbamates may cause a significant reduction in non-target organisms. Carbamates are toxic for worms and other organisms living in the soil. Although a great reduction in the earthworm population may occur when applying carbamates to the soil, numbers will return to normal, because of the rather rapid breakdown of these compounds.

In general, the toxicity of carbamates for wildlife is low, but exceptions exist. This means that, in order to judge the impact of carbamates on the organisms in the environment, the information on the individual carbamates should be referred to. As a rule, birds are not very sensitive to carbamates while bees are extremely sensitive.

1.1.6 Kinetics and metabolism

The metabolic fate of carbamates is basically the same in plants, insects, and mammals. Carbamates are usually easily absorbed through the skin, mucous membranes, and respiratory and gastrointestinal tracts, but there are exceptions. Generally, the metabolites are less toxic than the parent compounds. However, in certain cases, the metabolites are just as toxic or even more toxic than the parent carbamate. In most mammals, the metabolites are mainly excreted rather rapidly in the urine. The dog seems to be different in this respect. Accumulation takes place in certain cases, but is of minor importance because of the rapid metabolism.

The first step in the metabolism of carbamates is hydrolysis to carbamic acid, which decomposes to carbon dioxide (CO_2) and the corresponding amine.

The mechanism of hydrolysis is different for N-methyl and N-dimethyl derivatives. The N-methyl carbamates pass through an isocyanate intermediate, whereas in the hydrolysis of N-dimethylcarbamates, an addition product with a hydroxyl ion is formed yielding the alcohol and N-dimethyl substituted acid. The rate of hydrolysis by esterases is faster in mammals than in plants and insects.

Apart from hydrolysis, oxidation also takes place including: hydroxylation of the aromatic ring, O-dealkylation, N-methyl hydroxylation, N-dealkylation, oxidation of aliphatic side chains, and sulfoxidation to the corresponding sulfone. Oxidation is associated with the mixed-function oxidase (MFO) enzymes. Conjugation leads to the formation of O- and N-glucuronides, sulfates, and mercapturic acid derivatives in mammals. Glycosides and phosphates are conjugation products more common in plants.

The metabolism of a number of carbamates is discussed in the text.

Little information is available on the distribution of carbamates in the various organs and tissues in mammals following exposure by inhalation or the oral route. The organs in which residues have been reported are the liver, kidneys, brain, fat, and muscle. The half-life in the rat is of the order of 3 - 8 h. From the limited data available, it seems that the excretion of carbamates via urine is also rapid in man, and that the metabolic pathways in man are the same as those in the rat.

1.1.7 Mechanism of toxicity

Carbamates are effective insecticides by virtue of their ability to inhibit acetylcholinesterase (AChE) (EC 3.1.1.7) in the nervous system. They can also inhibit other esterases.

The carbamylation of the enzyme is unstable, and the regeneration of AChE is relatively rapid compared with that from a phosphorylated enzyme. Thus, carbamate pesticides are less dangerous with regard to human exposure than organophosphorus pesticides. The ratio between the dose required to produce death and the dose required to produce minimum symptoms of poisoning is substantially larger for carbamate compounds than for organophosphorus compounds.

Because of their chemical structure, carbamates do not cause delayed neuropathy.

1.1.8 Effects on experimental animals and in vitro test systems

The acute toxicity of the different carbamates ranges from highly toxic to only slightly toxic or practically non-toxic.

The LD_{50} for the rat ranges from less than 1 mg/kg to over 5000 mg/kg body weight. For certain methyl carbamates, the LD_{50} is 20 or more times the corresponding ED_{50}. This means that, in general, an early indication of poisoning can be obtained before a lethal dose is absorbed.

A dose-effect relationship exists between the dose, the severity of symptoms, and the degree of cholinesterase (ChE) inhibition. Because most carbamates have a low volatility, inhalation studies are mainly carried out using a dust or mist. In these studies, the toxicity is highly dependent on the size of the particles or droplets and, therefore, difficult to evaluate.

The acute dermal toxicity of carbamates is generally low to moderate; an exception is aldicarb, which is highly toxic. It should be noted that data are available for only a limited number of substances.

Carbamates produce slight to moderate skin and eye irritation, depending on the vehicle used, duration of contact, and on whether the substance is applied to the abraided or intact skin. From the available data, it cannot be excluded that some of the carbamates will have a slight to moderate sensitization potential.

Short- and long-term toxicity studies have been carried out. Some carbamates are very toxic and others are less toxic in long-term studies. From these studies, it is evident that, apart from the anticholinesterase activity, the following changes can be found: an influence on the haemopoietic system, an influence on the functioning of, and, at higher dosages, degeneration of, the liver and kidneys, and degeneration of testes. These abnormalities in the different organ systems depend on the animal strain and on the chemical structure of the carbamate. A clear influence on the nervous system, functional as well as histological, was found, particularly in non-laboratory animals such as pigs.

For many years, long-term toxicity data on carbamates have been evaluated by the FAO/WHO Joint Meeting on Pesticide Residues (JMPR), and a number of ADIs for carbamates have been established. In Annex II and III, the no-observed-adverse-effect levels and the ADIs are summarized.

A considerable number of reproduction and teratogenicity studies have been carried out with different carbamates and various animal species. Different types of abnormalities were found, i.e., increase in mortality, disturbance of the endocrine system, and effects on the hypophysis and its gonadotrophic function. These effects were mainly seen at high dose levels. Generally, the fetal effects included an increase in mortality, decreased weight gain in the first weeks after birth, and induction of early embryonic death. All these effects can be summarized as embryotoxic effects.

Certain carbamates also induce teratogenic effects, mainly at high dose levels applied by stomach tube. When the same dose level was administered with the diet, no effects were seen.

1.1.9 Mutagenicity and related end-points

The well-known carbamates have been tested for their mutagenic activity in different test systems. Some induce mutagenic effects, others are negative. In general, the methyl carbamates are negative in mammalian tests, while compounds such as carbendazim, benomyl, and the 2 thiophanate derivatives showed a positive effect with very high dose levels in certain systems. The benzimidazole moiety may act as a base analogue for DNA and as a spindle poison. They are antimitotic agents and cause mitotic arrest, mitotic delay, and a low incidence of chromosome damage. Sometimes, the results are contradictory or cannot be reproduced, but positive results for point mutation and chromosome aberrations are well documented. These benzimidazole derivatives can be considered as weak mutagenic compounds.

1.1.10 Carcinogenicity

Ethyl carbamate (urethane) is a well-known carcinogen, and it seems that its chemical structure is optimal for such an effect. Any change in the molecule seems to decrease the carcinogenic potency, particularly when the ethyl group is replaced by larger side chains. Alkyl groups on the nitrogen also reduce this activity.

However, no clear indications of carcinogenic effects have been found in the available long-term carcinogenicity studies with different carbamates.

The carcinogenicity studies with benzimidazole derivatives showed either positive or equivocal results. Added to the fact that certain mutagenicity studies also give positive results, it cannot be excluded that these compounds may have carcinogenic or promotor properties. It should be kept in mind that the dose levels in most tests were of the order of 50 and 500 mg/kg body weight.

Carbamate pesticides may be converted to *N*-nitroso compounds. This was demonstrated in a great number of *in vivo* nitrosation studies in which high levels of the carbamates were administered to animals in combination with high levels of nitrite. These *N*-nitroso compounds have to be considered as mutagenic and carcinogenic. However, the amount of nitroso compounds that can be expected to result from dietary intake of carbamate pesticide residues is negligible in comparison with nitroso-precursors that occur naturally in food and drinking-water.

1.1.11 Effects on man

Health hazards for man occur mainly from occupational over-exposure to carbamate insecticides resulting in poisoning characterized by cholinergic symptoms caused by inhibition of the enzyme AChE. Various cases of intoxication have been described. Most of them were spraymen applying insecticides inside houses in the tropics to control mosquito vectors of malaria, or plant protection workers. The main routes of exposure are inhalation and skin.

From controlled human studies, it is clear that poisoning symptoms can be seen a few minutes after exposure, and can last for a few hours. Thereafter, recovery starts and within hours, the symptoms disappear, and the ChE activity in erythrocytes and plasma returns to normal, because the carbamate is rather rapidly metabolized and the metabolites excreted. The appearance of these metabolites in the urine may be used for biological monitoring. Apart from the symptoms indicative of ChE poisoning, other signs and symptoms induced by certain carbamates have been described, such as skin and eye irritation, hyperpigmentation, and influence on the function of testes (slight increase of sperm abnormalities). These signs and symptoms were found in a few studies and should be confirmed before it can be stated that they were induced by carbamates. Epidemiological studies with persons primarily exposed to carbamates are not available.

1.1.12 Previous evaluations by international bodies

Previous evaluations of individual carbamates by the Joint FAO/WHO Meeting on Pesticide Residues (JMPR) and the International Agency for Research on Cancer (IARC) are summarized in Annex III. The WHO recommended classification of pesticides by hazard is included, and the availability is indicated of WHO/FAO Data Sheets and the IRPTC data profile and legal file on the substance.

1.2 Recommendations

1. More up-to-date information is necessary on the world-wide production and uses of the different carbamates.

2. Except for the well-known carbamates, such as carbaryl, benomyl, and carbendazim, more information is required on environmental pathways, concentrations, and distribution.

3. More information is necessary on the occurrence and fate of the carbamates in surface water, soil, and groundwater, and their impact on plants, invertebrates, and mammals.

4. Further studies are necessary on the occurrence of carbamates in the different food chains (bioaccumulation), and in food and drinking-water (market-basket or total-diet studies), in order to estimate the daily exposure of the human population.

5. More information is needed, for certain carbamates, on the acute and long-term toxicity in aquatic and terrestrial organisms.

6. In general, data on the mode of action and metabolism are available; however, more knowledge is needed on the distribution of the carbamates in organs and tissues in mammals.

7. For many carbamates, information concerning long-term toxicity, mutagenicity, carcinogenicity, reproduction, and teratogenicity is still needed. This is the reason why acceptable daily intakes have not been established for two-thirds of the carbamates.

8. Apart from a number of studies on human volunteers and a number of accidents with well-known carbamates, information on the effects of human exposure to carbamates is missing. There are no epidemiological studies. More information should be collected to evaluate the health risk in cases of human exposure to carbamates. In particular, more information is needed to elucidate the mechanisms of the effects of benzimidazole carbamates on the testes.

9. More data are needed on the use of oximes in the therapy of poisoning, especially in light of the fact that there are unconfirmed reports that oximes might potentiate carbamate toxicity.

10. Further work should be done to develop more adequate analytical methods (i.e., faster procedures and simpler equipment) to determine carbamate residues in biological material (urine, blood, and food).

11. Information is required concerning the changes in toxicity due to impurities that can arise in pesticides as a consequence of different manufacturing processes, formulating practices, and improper storage.

12. Users should be encouraged to be aware of the necessity to establish safe re-entry periods according to local conditions.

Recommendations for further work have also been described in previous monographs on carbamates published in the WHO Pesticide Residues Series and in the FAO Plant Production and Protection Papers, as a result of the Joint FAO/WHO Meetings on Pesticide Residues (JMPR).

2. IDENTITY, PHYSICAL AND CHEMICAL PROPERTIES, ANALYTICAL METHODS

2.1 Identity

Carbamates are *N*-substituted esters of carbamic acid. Their general formula is:

$$R^1NH - \overset{\overset{\displaystyle O}{\|}}{C} - OR^2$$

where R^2 is an aromatic or aliphatic moiety. Three main classes of carbamate pesticides are known:

(a) carbamate insecticides; R^1 is a methyl group;

(b) carbamate herbicides; R^1 is an aromatic moiety; and

(c) carbamate fungicides; R^1 is a benzimidazole moiety.

2.2 Physical and Chemical Properties

In general, simple esters or *N*-substituted derivatives of carbamic acid are unstable compounds, especially under alkaline conditions. Decomposition takes place, and the parent alcohol, phenol, ammonia, amine, and carbon dioxide are formed.

The salts and esters of substituted carbamic acid are more stable than carbamic acid. This enhanced stability is the basis for the synthesis of many derivatives that are biologically active pesticides.

Carbamate ester derivatives are crystalline solids of low vapour pressure with variable, but usually low, water solubility. They are moderately soluble in solvents such as benzene, toluene, xylene, chloroform, dichloromethane, and 1,2-dichloroethane. In general, they are poorly soluble in nonpolar organic solvents such as petroleum hydrocarbons but highly soluble in polar organic solvents such as methanol, ethanol, acetone, dimethylformamide, etc.

The carbamate derivatives with herbicidal action (such as pyrolan and dimetilan) are substantially more stable to alkaline hydrolysis than the methyl carbamate derivatives (carbaryl and propoxur), which have an insecticidal action. For example, the half-life of carbaryl is 15 min at pH 10 compared with 10 days at pH 7. However, pyrolan and dimetilan do not hydrolyse in the pH range of 4 - 10 (Aly & El Dib, 1972). Instability with alkali is of use for decontamination and clean-up. Vassilieff & Ecobichon (1982) showed that, in acid fresh water, which is characteristic of the lakes and streams in heavily forested Canada, aminocarb would be rather stable and would persist long enough to be bioaccumulated by various trophic levels of food chains.

The carbamate fungicides carbendazim, benomyl, and thiophanates are related. Carbendazim and benomyl are derivatives of benzimidazole. Carbendazim is slowly hydrolysed by alkali to 2-aminobenzimidazole, but it is stable as acid-forming water-soluble salts (Sypesteijn et al., 1977). Benomyl breaks down to methyl 2-benzimidazole carbamate (MBC) in water. Benomyl is rather unstable in common solvents (White et al., 1973; Chiba & Doornbos, 1974).

The names, chemical structure, and pesticidal activity of the principal carbamates are presented in Table 1, and the CAS number, chemical name, common names, molecular formula, relative molecular mass, and selected chemical and physical properties are summarized in Annex I. It should be noted that it is not within the scope of this general introduction to describe all the physical and chemical properties of each pesticide in detail.

2.3 Analytical Methods

Analysis for pesticide residues consists of sampling the contaminated environmental material or matrix, extracting the pesticide residue, removing interfering substances from the extract, and identifying and quantifying the pesticide contaminant. The manner in which the matrix material is sampled, stored, and handled can affect the results; samples should be truly representative, and their handling and storage must not further contaminate or degrade the contaminant to be measured. Many detection methods are available, and the one chosen depends on the physical and chemical properties of the contaminant as well as on the equipment available.

A detailed review of all the analytical procedures to determine carbamates in the different matrices is beyond the scope of this document. However, for better understanding of the validity of the data, a brief summary of some of the analytical procedures for different carbamates is included in Table 2.

Enzymatic methods to determine erythrocyte- and plasma-ChE activity are used as monitors of exposure and systemic absorption of organophosphorus compounds and carbamate pesticides. In the case of carbamates, the inhibition may not be easily detected because of the rapid reversibility of the carbamate-enzyme inhibition reaction.

Conditions and time of storage must be carefully controlled before measurement of activity. Methods must be

Table 1. Relationship of chemical structure and pesticidal activity of carbamates

Pesticidal activity	Chemical structure	Common or other names
Insecticide	$CH_3\text{-NH-}\overset{O}{\overset{\|}{C}}\text{-O-aryl}$	aldoxycarb, allyxycarb, aminocarb, BPMC, bendiocarb, bufencarb, butacarb, carbanolate, carbaryl, carbofuran, cloethocarb, dimetilan, dioxacarb, ethiofencarb, formetanate, hoppcide, isoprocarb, trimethacarb, MPMC, methiocarb, metolcarb, mexacarbate, pirimicarb, promacyl, promecarb, propoxur, MTMC, XMC, xylylcarb
	$CH_3\text{-NH-}\overset{O}{\overset{\|}{C}}\text{-O-N-alkyl}$	aldicarb, methomyl, oxamyl, thiofanox, thiodicarb
Herbicide	$\text{aryl-NH-}\overset{O}{\overset{\|}{C}}\text{-O-alkyl}$	asulam, barban, carbetamide, chlorbufam, desmedipham, phenmedipham, swep
	$\text{alkyl-NH-}\overset{O}{\overset{\|}{C}}\text{-O-aryl}$	dichlormate, karbutilate, terbucarb
Herbicide and sprout inhibitors	$\text{aryl-NH-}\overset{O}{\overset{\|}{C}}\text{-O-alkyl}$	propham, chlorpropham
Fungicide	$\text{aryl-NH-}\overset{O}{\overset{\|}{C}}\text{-O-alkyl}$	benomyl, carbendazim, thiophanate-methyl, thiophanate-ethyl

Table 2. Analytical methods for carbamate pesticide residues

Chemical	Sample type	Extraction and clean-up	Method of detection[a]	Detection[b] limit	Reference
aldicarb	cotton seed, fruit/vegetables	acetone-water-peracetic acid extraction evaporation, absorption on Florisil column, elution with acetone-ether	GLC/SFPD	0.01 mg/kg	Romine (1973)
asulam	crops	acetone extraction, absorption on Florisil column, elution, reaction with chromophore	colorimetry	0.05 mg/kg	Brockelsby & Muggleton (1973)
benomyl	soils, fruit/vegetables, tissues	acidic methanol extraction and ethyl acetate extraction, conversion to carbendazim	HPLC/UV	0.05 mg/kg	Bleidner et al. (1978)
	soils	acetone-ammonium chloride extraction and solvent partition	UV		Austin & Briggs (1976)
	soils, fruit/vegetables, tissues	review of methods including those for carbendazim and thiophanate-methyl		0.05 - 3.0 mg/kg	Baker & Hoodless (1974)
carbendazim	water	reverse phase and adsorption systems	HPLC, variable wavelength UV-detector		Austin et al. (1976)
carbaryl	urine	acid hydrolysis, benzene extraction, reaction with chloroacetic anhydride, absorption on silica gel column and elution with benzene-hexane	GLC/ECD	0.02 mg/litre	Shafik et al. (1971)
carbaryl	plant tissue	extraction, hydrolysis, and reaction with chromophore	colorimetry	0.1 mg/kg	Stansbury & Miskus (1964)

Table 2 (contd).

Chemical	Sample type	Extraction and clean-up	Method of detection[a]	Detection[b] limit	Reference
methiocarb	fruit and vegetables	acetone extraction of carbamates and free phenols, partition on silica gel column, hydrolysis of conjugated phenols	GLC/SFPD	0.01 mg/kg	Bowman & Beroza (1969)
methomyl		ethyl acetate extraction, hexane extraction, chloroform solution, hydrolysis and ethyl acetate extraction	GLC/SFPD	0.02 mg/kg	Leitch & Pease (1973)
phenmedipham	plant tissues	hydrolysis, distillation-extraction, and reaction with chromophore	colorimetry	0.05 mg/kg	Kossmann & Jenny (1973)
propoxur	plant and animal tissues	acetone-chloroform extraction, absorption on Florisil column, chloroform elution, solution in benzene,	GLC/ECD	0.1 mg/kg	Anderson (1973)
	milk	and reaction with trichloroacetyl chloride		0.01 mg/kg	
thiofanox	soil	acetone extraction, oxidation with hydrogen peroxide, absorption on Florisil column and elution with chloroform and diethyl ether, solution benzene	GLC/SFPD		Chin et al. (1975)

a GLC = gas-liquid chromatography.
ECD = electron-capture detector.
SFPD = sulfur (sensitive) flame photometric detector.
HPLC = high-performance liquid chromatography.
UV = ultraviolet.

b Limit of detection is the sensitivity of the method; however, each method may not be able to measure all the contaminants originally present in the sample, i.e., the recovery rates for spiked samples < 100%.

chosen that allow a short time for hydrolysis of the substrate (Ellman et al., 1961; Voss & Schuler, 1967; Wilhelm & Reiner, 1973; Wilhelm et al., 1973; Reiner et al., 1974; Abd-Elroaf et al., 1977; Izmirova, 1980).

A colorimetric screening method has been described, to estimate Unden®, carbofuran, and carbaryl in the air of the manufacturing and formulating plants, as well as in washings from body surfaces, hands, and other contaminated surfaces (Izmirova, 1980; Izmirova & Izmirov, unpublished report, 1985)[a].

---

[a] Pharmatest-cholinesterase reactive papers for determination of cholinesterase activity (ChEA) of serum or plasma.

3. SOURCES OF HUMAN AND ENVIRONMENTAL EXPOSURE

3.1 Natural Occurrence

Methyl carbamates are related to a naturally-occurring carbamate alkaloid, physostigmine, isolated from the calabar bean (*Physostigma venenosum*) in 1864.

Physostigmine (or eserine) has a pronounced cholinergic action (Still & Herrett, 1975).

3.2 Man-Made Sources

Carbamate pesticides have been produced and in commercial use since the 1950s.

Benomyl and carbendazim, belonging to the benzimidazole group of fungicides, came on the market around 1970.

3.2.1 Production levels, processes, and uses

The carbamates included in this review are those mainly used in agriculture. They are some of the many synthetic organic pesticides that have been produced on a large scale. Additional uses are as biocides for industrial or other commercial applications and in household products; a potential use is in public health vector control.

Global consumption of the carbamate insecticides and herbicides over the period 1974-82 is summarized in Table 3. The data are not complete but give an estimate of the magnitude of the consumption and distribution of the compounds throughout the world. The reported global consumption is between 20 000 and 35 000 tonnes/year. The herbicidal carbamates, an integral part of industrialized agriculture, are used mainly in North and Central America, and Europe, with little reported use in Africa, Asia, and South America. However, in the period mentioned, carbamate insecticides were substantially used in Asia (FAO, 1985).

Table 3. Global consumption of carbamate insecticides (in 100 kg)[a]

Region	1974-76	1981	1982	1983
Africa				
Sierra Leone	25			
Sudan	1590			
Zimbabwe		4837		
North/Central America				
Bermuda				
Canada	7911			
Cuba	7333			
El Salvador	507			
Mexico	25 367	21 630	21 380	17 210
Montserrat		2	2	
USA		125 000	115 000	
South America				
Argentina		2960	2700	
Guyana			36	
Suriname		369		
Uruguay	63	74	70	179
Asia				
Brunei		7	12	29
Cyprus	20	124	21	
Hong Kong	100	125	64	102
India	33 447	32 160	23 730	
Israel	1333	2160	2580	2270
Japan		27 770		
Jordan		2000	16 745	
Korea, Republic	7618	19 270	17 716	
Kuwait	2			
Oman		134	6	16
Pakistan	298	3511	1136	
Saudi Arabia	63			
Turkey		635	666	
Europe				
Austria	213	184	153	86
Czechoslovakia	1490	569	712	
Denmark	60		307	440
Finland	7			
Greece	8767			
Hungary	5854	1539	3396	11 650
Italy	28 014	28 284	23 041	
Norway		5	5	10
Poland	12 353	3325	4365	4977
Portugal	512	216	200	
Sweden		94	130	
Switzerland	133	140	120	

[a] From: FAO (1985).

4. ENVIRONMENTAL TRANSPORT, DISTRIBUTION, AND TRANSFORMATION

4.1 Transport and Distribution Between Media

4.1.1 Air

In general, the vapour pressure of carbamates is rather low. Some of them may sublimate slowly at room temperature, and this may also explain their loss from soil surfaces (Gray, 1971).

4.1.2 Water

The aqueous environment is an important factor in transport. Carbamates may enter surface water from industrial wastes, accidental spillage, and dumping. However, the hazard of this is limited by their rapid decomposition under aqueous conditions. Thus, while long-term contamination by this type of compound is unlikely, adverse effects on aquatic animals may result from direct addition or from run-off shortly after application.

4.1.3 Soil

Reviews of the action of herbicidal carbamates in the soil have been made by Gray (1971) and Ashton & Crafts (1973).

Herbicidal carbamates are often applied directly to the soil, whereas insecticidal carbamates, generally applied to plants, reach the soil either directly or indirectly. Several factors are involved in the degradation of carbamates in the soil. These include volatility, leaching, soil moisture, absorption, pH, temperature, photodecomposition, microbial degradation, and soil type (Ogle & Warren, 1954).

Different soil types possess different binding capabilities. In general, carbamate insecticides are not very persistent in the soil. However, the fungicide carbendazim is very persistent, with a half-life of about one year.

Because of the many factors involved and the fact that many carbamates have different properties, it is clear that results with one soil type and one carbamate cannot be extrapolated to others.

In general, the water solubility of carbamates is rather low and may explain the relative immobility of the carbamate herbicides in the soil with regard to leaching and diffusion. When applied to the soil, some chloropropham (CIPC) can be lost by vaporization or sublimation, but otherwise it is tightly adsorbed to certain soils (Ogle & Warren, 1954). It was found by De Rose (1951) that chloropropham (CIPC) persisted in the soil about twice as long as propham (IPC). The disappearance of propham from the soil took about 24 days, while chloropropham persisted at least 48 days. Water can displace the carbamates from adsorption sites and cause them to be lost by volatilization (Parochetti & Warren, 1966).

The degradation of the carbamate herbicides in the environment is summarized in Fig. 1, using chloropropham as a representative example of the group.

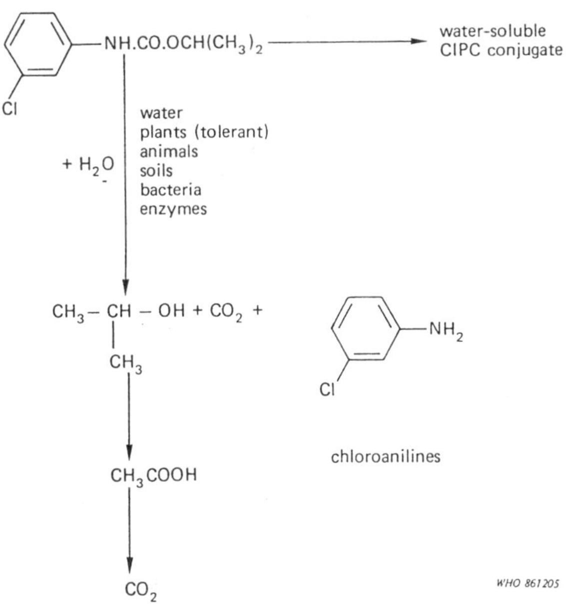

Fig. 1. Chloropropham (CIPC) degradation as an example of the breakdown of carbamate herbicides in the environment.

Soil run-off and leaching characteristics of benomyl and its metabolites were studied in the greenhouse and laboratory using ^{14}C-labelled materials on soil and turf plots and on soil thin-layer chromatographic plates (Rhodes & Long, 1974). The studies showed that benomyl and its metabolites are immobile in soil and do not leach or move from the site of application.

The first step in the metabolic degradation of these carbamates in soil is hydrolysis (Sonawane & Knowles, 1971). Simple esters are hydrolysed to the parent (unstable) acid and alcohol. The general reaction of carbamates and the breakdown of carbamates in soils are described in detail in the review of Still & Herrett (1975). Chloropropham, barban, and swep are hydrolysed to their respective anilines (Still & Herrett, 1975). In alkaline soil, phenmedipharm converts hydrolytically to methyl-3-hydroxyphenyl-carbamate, which in turn hydrolyses to 3-aminophenol (Sonawane & Knowles, 1971). The substituted anilines that are formed are further metabolized by oxidation processes (Kaufman & Blake, 1973; Still & Herrett, 1975).

Methomyl and oxamyl degrade rapidly with carbon dioxide as the end product (Harvey & Pease, 1973; Harvey & Han, 1978b).

In soils, the predominant metabolic pathway is cleavage of the carbamate bond to yield the alcohol and amino moieties, which may be further metabolized by the soil-plant system.

Williams et al. (1976a,b) found that the degradation of carbofuran in British Columbia was slower than expected. Soils showed rather high residues of this compound (up to approximately 4 mg/kg soil), and there was some evidence of build-up where treatments were repeated for two successive years. While carbofuran has a half-life of up to 50 weeks in neutral or acid soils, it degrades rapidly in alkaline soils.

4.1.4 Vegetation and wildlife

Plant metabolism of carbanilate herbicides has been shown to involve aryl hydroxylation and conjugation besides hydrolytic breakdown. Consequently, they are readily transformed from lyophilic compounds to hydrophylic metabolites, which contain the intact carbamoyl group. These polar products are not translocated but remain in the plant at the site of formation. All available data indicate that the hydroxylated carbamate metabolites are detoxification products (Still & Herrett, 1975). In studies on chloropropham in oat shoots, Still & Rusness (1977) found that the phenolic metabolites were converted to an S-cysteinyl conjugate, i.e., S-cysteinyl-hydroxychloropropham.

A large number of studies on the absorption and translocation of carbamate herbicides have been carried out to study the fate of these compounds in plants. The results suggest that plant leaf surfaces are a barrier to the absorption of carbamates. Roots, however, absorb the herbicides to a much greater extent, and the carbamate moves to all plant parts. Thus, it is suggested that carbamates are exclusively distributed via the apoplastic system. The carrier that is used plays an important role in these absorption studies.

Many studies are available concerning the actual entry of carbamates into the plant, their hydrolysis, aromatic ring hydroxylation, hydrolytic breakdown, and oxidation of aliphatic groups. Other types of metabolic reactions in different plant species following different methods of application have also been described (Baldwin et al., 1954; Abdel-Wahab et al., 1966; Prendeville et al., 1968; Knowles & Sonawane, 1972; Still & Mansager, 1972, 1973a,b, 1975; Wiedmann et al., 1976; Guardigli et al., 1977).

For the metabolic pathway in wildlife, see sections 6 and 7.

4.1.5 Entry into the food chain

Carbamates are metabolized or broken down in soil, plants, and animals. The questions of the environmental impact and the significance of the metabolites with respect to persistence and bioaccumulation in certain species and in the food chain have still to be answered. For the moment, it appears that most, if not all, of the known carbamate metabolites are biodegraded rapidly and are less toxic for the environment than the parent molecules (Still & Herrett, 1975).

4.2 Biotransformation

4.2.1 Microbial degradation

The carbamates are readily degraded by soil microorganisms in most soils (Gray, 1971). Environmental conditions that favour the growth and activity of microorganisms also favour degradation (Ogle & Warren, 1954). The residual herbicidal activity of both barban and propham persists longer in sterile soil than in non-sterile soil.

Kaufman & Kearney (1965) and Kaufman & Blake (1973) isolated (by soil-enrichment techniques) and identified a soil microorganism capable of degrading carbamates such as chlorpropham. They suggested that hydrolysis was a major degradation pathway in soils. Each of the soil microorganisms demonstrated a different range of substrate specificity, but all were capable of degrading and dehalogenating a variety of pesticides (production of 3-chloroaniline and subsequent liberation of free chloride ion).

These studies illustrated the tremendous adaptability of the soil microbial populations in altering the persistence and character of the different foreign compounds. Studies carried out by Clark & Wright (1970) confirmed that cultures of *Arthrobacter* sp. and *Achromobacter* sp. were able to convert phenyl carbamates to the corresponding aniline compounds.

Suzuki & Takeda (1976) studied the metabolism of dimetilan in Aspergillus niger van Tieghem. Together with hydrolysis, oxidation of the alkyl side chain appeared the most important modification (detoxification) process. Williams et al. (1976b) found that carbofuran was rapidly degraded in soils containing high levels of actinomycetes.

4.2.2 Photodegradation

N-methyl carbamates absorb radiation available in the solar region (λ = 300 nm) and hence, would be expected to undergo photo-oxidation as well as metabolic degradation. Addison et al. (1974) studied the fate of various *N*-methyl carbamates when solutions were sprayed on bean foliage and exposed to sunlight or artificial light (λ = 254 nm). It is not entirely clear whether the products they observed resulted solely from a photochemical reaction or from absorption followed by enzymatic attack (Addison et al., 1973, 1974).

4.2.3 Photodecomposition in the aquatic environment

Carbamate insecticides in water are subject to photodecomposition under the effects of ultraviolet radiation (UVR). The pH of the aqueous medium was found to be an important factor in relationship to the rates of photolysis of carbaryl and propoxur, which were slow at low pH values and tended to increase with increaseing pH value. However, the decomposition of, for instance, dimetilan, was not affected by the pH of the irradiated medium. The primary effect of the UVR appears to be cleavage of the ester bond resulting in the production of the phenol or heterocyclic enol of the carbamate esters tested. The hydrolysis products produced were further photodecomposed to other unidentified degradation products. Carbaryl produced 5 degradation products, one of which was identified as 1-naphthol. It is assumed that, apart from the cleavage of the ester bond, changes at other positions in the molecule are produced by UVR. However, the intact carbamate ester group is retained, and consequently the ChE-inhibiting activity (Crosby et al., 1965; Crosby 1969; Aly & El Dib, 1972).

Similar results were reported for the irradiation products of other carbamate esters such as aminocarb, mexacarbate, and a number of analogues (Eberle & Gunther, 1965; Abdel-Wahab F & Casida, 1967).

When the half-life for photodecomposition was studied in a number of carbamate insecticides, the results suggested that the light absorption characteristics of the insecticide influenced the extent of its photodecomposition by a specific light source. However, the extent of photodecomposition was not the same under different conditions of irradiation (presence of sol- vents) and wave-length (Crosby et al., 1965; Eberle & Gunther, 1965).

Thus, it seems reasonable to suggest that photodecomposition may account for some loss of carbamate insecticides in clear surface waters exposed to sunlight for a long period. However, photolysis may be a minor factor in the decomposition of these compounds in highly-turbid waters, where the penetration of light will be greatly reduced (Aly & El-Dib, 1971, 1972).

5. ENVIRONMENTAL LEVELS AND HUMAN EXPOSURE

5.1 Exposure of the General Population

5.1.1 Food and drinking-water

From a market-basket study conducted during 1977-80 by the Finnish National Board of Trade and Consumer Interests, it became apparent that the residues of benomyl in food had increased. During the period 1974-76, the average benomyl intake was about 0.2 mg/person per year from both domestic and imported foods. During the period 1977-80, average intake was 9 mg/year. A later more accurate study showed an intake of 14 mg/person per year (Finnish National Board of Health, 1982).

Contamination of groundwater and drinking-water sources by aldicarb, a carbamate that is used as an insecticide and as a nematocide, was reported from the USA in 1979 and 1980-81; levels even higher than 0.075 mg/litre were found (Rothschild et al., 1982; Zaki et al., 1982). In the Federal Republic of Germany, detectable amounts of up to 0.001 mg aldicarb/litre were found in a few samples of drinking-water (Federal Republic of Germany, personal communication, 1985)[a].

---

[a] Letter to the IPCS from the Ministry of the Interior of the Federal Republic of Germany, Bonn, Federal Republic of Germany, 28 May, 1985.

6. METABOLISM AND MODE OF ACTION

Most carbamates are active inhibitors of AChE and they do not require metabolic activation. However, some carbamates, such as the benzimidazole carbamates, do not have anticholinesterase activity. Carbamates undergo metabolism, and the metabolites are generally less toxic than the parent compound; some exceptions will be discussed later. The metabo- lism of carbamates is basically the same in mammals, insects, and plants, and the toxic effects of carbamates are similar in mammals and insects. Carbamates do not accumulate in the mammalian body, but are rapidly excreted, mainly via the urine.

6.1 Mode of Action

Carbamates are effective insecticides by virtue of their ability to inhibit AChE in the nervous system. AChE catalyses the hydrolysis of the neurotransmitter acetylcholine (ACh) to choline and acetic acid.

ACh is the synaptic mediator of nerve impulses in the nervous system of mammals and insects:

(a) as a neurotransmitter in the brain of mammals, and in the central nervous system of insects;

(b) as a pre-ganglionic neurotransmitter in the autonomic nervous system of mammals;

(c) in post-ganglionic nerve endings of the autonomic nervous system; and

(d) at the neuromuscular junction of skeletal muscle.

Carbamates, like organophosphates, can inhibit esterases that have serine in their catalytic centre; these are called serine-esterases or beta-esterases. Although the inhibition of serine-esterases other than AChE is not significant for the toxicity of the compounds, it may have significance for the potentiation of toxicity of other compounds after long-term low-level exposure (Sakai & Matsumura, 1968, 1971; Aldridge & Magos, 1978).

To understand the mechanism of toxicity, it is necessary to examine the events that take place at the neuromuscular junction. When a muscle is innervated, as shown in Fig. 2, a nerve impulse moving down a neuron reaches the nerve ending where ACh, which is stored in vesicles at the nerve endings, is released into the junction. Within 2 - 3 ms, ACh impinges on the receptor side of the muscle.

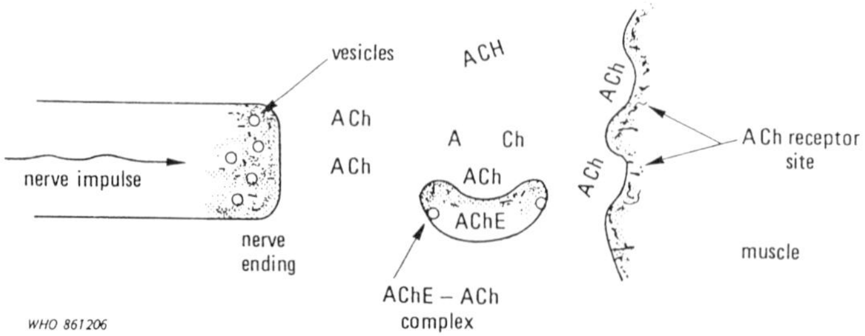

Fig. 2. Schematic drawing of events in neuromuscular function during stimulation.

AChE then hydrolytically converts the ACh into choline and acetic acid, which results in a decrease in the concentration of ACh and cessation of muscle contraction. When AChE is inhibited by a carbamate ester, it can no longer hydrolyse the ACh. Thus, the ACh concentration remains high in the junction giving rise to continuous stimulation of the muscle which, in turn, leads to exhaustion and tetany.

Thus, inhibition of AChE by carbamate esters, causes toxic effects in animals and human beings that result in a variety of poisoning symptoms and eventually culminate in respiratory failure and death.

The mechanism of inhibition of AChE by a carbamate ester can be formulated as follows (Reiner & Aldridge, 1967; Aldridge & Reiner, 1972):

$$\text{AChE} + \text{RO-}\overset{\overset{\text{O}}{\|}}{\text{C}}\text{-NHCH}_3 \underset{k_{-1}}{\overset{k_1}{\rightleftharpoons}} [\text{AChE.RO-}\overset{\overset{\text{O}}{\|}}{\text{C}}\text{-NHCH}_3] \text{ (enzyme-carbamate complex)}$$

$$\downarrow k_c$$

$$\text{AChE} + \text{CH}_3\text{NH}_2 + \text{CO}_2 \underset{\text{H}_2\text{O}}{\overset{k_r}{\longleftarrow}} \text{AChE-}\overset{\overset{\text{O}}{\|}}{\text{C}}\text{-NHCH}_3 + \text{RO}^-$$

(carbamylated enzyme)

where: k_1 = second order rate constant for formation of complex;

k_{-1} = first order rate constant for breakdown of complex to starting materials;

k_c = first order rate constant for carbamylation of the enzyme;

k_r = first order rate constant for hydrolysis of the carbamylated enzyme.

The site for carbamylation of the enzyme is the hydroxyl moiety of the serine amino acid. The rate of regeneration of the carbamylated enzyme to AChE (k_r) is relatively rapid compared with that of an enzyme that has been inhibited (phosphorylated) by an organophosphorus pesticide (Reiner & Aldridge, 1967; Reiner, 1971). Thus, human exposure to the carbamate pesticides is less dangerous than exposure to organophosphorus pesticides, because the ratio between the dose required to produce mortality and the dose required to produce minimum poisoning symptoms is, in general, substantially larger for carbamate compounds than for organophosphorus compounds (Goldberg et al., 1963; Vandekar, 1965; Vandekar et al., 1971). An individual experiencing symptoms of poisoning from a carbamate is more likely to recover on termination of exposure and with appropriate medical treatment than an individual poisoned by an organophosphorus compound.

The spontaneous reactivation of various carbamylated ChEs expressed as half-life, at pH 7.0 - 7.4 and 25 °C ranged between 2 and 240 min for AChE and between 2 and 17 min for serum-ChE. This instability of the carbamylated enzyme affects the determination of the inhibitory power of carbamates, the recovery after poisoning, and the determination of inhibition of blood-ChEs (Reiner, 1971).

6.1.1 Neuropathy target esterase

This esterase, formerly known as neurotoxic esterase (NTE), is the target for the initiation of delayed neuropathy caused by some organophosphorus esters. Both organophosphorylation of NTE and a subsequent "aging" reaction of the inhibited enzyme are required to initiate neuropathy. Certain *N*-aryl carbamates can also inhibit NTE, but the chemistry of such carbamates is such that no "aging" reaction is possible. These carbamate inhibitors do not initiate delayed neuropathy *in vivo* but can actually prevent the neuropathic effects of a challenge dose of an organophosphate given while the tissue NTE is carbamylated (Johnson, 1970). No delayed neuropathic effects have been observed in tests on carbamate pesticides according to protocols that detect neuropathic organophosphates, and this agrees with the mechanism described above.

6.2 Metabolism

The metabolic fate of carbamates is basically the same in plants, insects, and mammals.

Carbamates can penetrate the skin, mucous membranes, respiratory tract, and gastrointestinal tract of mammals.

6.2.1 Absorption

In vivo studies have shown that carbamates are almost completely absorbed during normal transit through the gastrointestinal tract.

According to absorption studies on rats, the dermal absorption of radioactive labelled benomyl is slow. After absorption, rapid metabolism and elimination via urine took place. After 1 h, small amounts of the major metabolites of benomyl, 5-hydroxy-2-benzimidazole carbamate (5-HBC), and methyl 2-benzimidazole carbamate (MBC), could be detected in urine, confirming the rapid metabolism of benomyl. No radioactivity was found in any body tissues sampled 24 h after application (FAO/WHO, 1985a).

6.2.2 Distribution

Ahdaya et al. (1981) studied the absorption and distribution of carbamates. They found that, in mice, all carbamates were rapidly distributed to the tissues and organs. The half-life values of penetration ranged from 8 to 17 min for carbamates.

No residues of benomyl or its degradation products were detected (level of sensitivity 0.02 mg/kg) in eggs from chickens fed 5 mg benomyl/kg diet. Only 5-HBC (0.03 - 0.06 mg/kg) was found in eggs from chickens fed 25 mg benomyl/kg for four weeks. No residues of benomyl or MBC were found in milk (< 0.02 mg/litre) from dairy cows fed benomyl for 32 days at dietary levels of 0, 2, 10, and 50 mg/kg. At the highest dietary level, the combined residue of 5-HBC plus 5-hydroxy-2-benzimidazole carbamate (4-HBC) in the milk was approximately 0.1 mg/litre. These metabolic studies appear to indicate that benomyl and its metabolites do not accumulate in animal tissues or animal products (Gardiner et al., 1974; FAO/WHO, 1985a).

A rat was administered a diet containing 2500 mg non-labelled benomyl/kg. Twelve days later, the animal received an intragastric intubation of 2-^{14}C-benomyl (7.7 mg). Urine and faeces were collected, and organs were analysed for radioactivity. The elimination was rather rapid and mainly via the urine. After 72 h, 0.2% of the administered dose was found in the liver, while < 0.01% was present in all the other organs and the carcass. The same results were found when a rat was administered 2-^{14}C-MBC (Gardiner et al., 1974)

A male beagle dog fed a diet containing non-labelled benomyl at 2500 mg/kg, and administered 30.8 mg 2-^{14}C-benomyl by capsule 7 days later, showed a completely different excretion pattern. After 3 days, residues were found only in the liver (0.31% of the administered dose); the levels in the other organs were < 0.01% (Gardiner et al., 1974).

Rats administered an oral dose of ^{14}C-labelled propham or chloropropham (labelled in the chain or in the ring), showed radioactivity in all tissues, with the highest concentration in the kidneys. The average biological half-life of ^{14}C from both compounds in most organs was short, ranging from 3 to 8 h. However, in brain, fat, and muscle, the half-life was about twice this value (Fang et al., 1974).

6.2.3 Metabolic transformation

6.2.3.1 Biotransformation mechanisms

Carbamate pesticides are transformed metabolically by a variety of chemical reactions into more water-soluble molecules with increased polar properties. The initial step, usually oxidative in nature, introduces a functional hydroxyl group that serves as a site for secondary conjugative reactions to yield products that can be excreted via the urine and/or faeces. In some cases, the oxygenated metabolites, such as 5-hydroxypropoxur and 5-hydroxycarbaryl, are known to be toxic and to possess anticholinesterase activity (Fig. 3) (Oonnithan & Casida, 1968; Black et al., 1973). This undoubtedly contributes to the overall toxicity of the parent compound (Black et al., 1973).

6.2.3.2 Oxidation

The principal route of metabolism of insecticidal carbamate esters is oxidative and is generally associated with the mixed-function oxidase (MFO) enzymes, which are present in several tissues. Examples of the sites of oxidative attack on a hypothetical methyl carbamate are given in Fig. 4.

Depending on the functional groups in the molecule, a variety of reactions catalysed by these enzymes may occur (Fukuto, 1972), as shown in Fig. 4. Typical oxidative reactions include: (a) hydroxylation of aromatic rings, or epoxidation; (b) O-dealkylation; (c) N-methyl hydroxylation; (d) N-dealkylation; (e) hydroxylation and subsequent oxidation of aliphatic side chains; and (f) thioether oxidation to sulfoxides and sulfones.

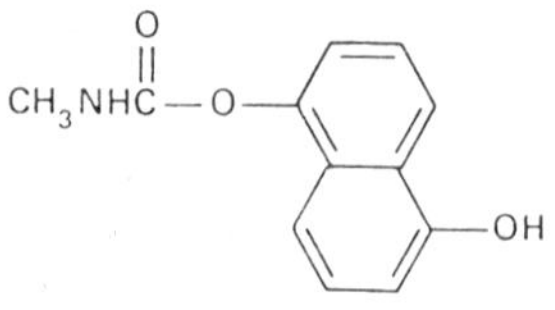

5 - hydroxy carbaryl

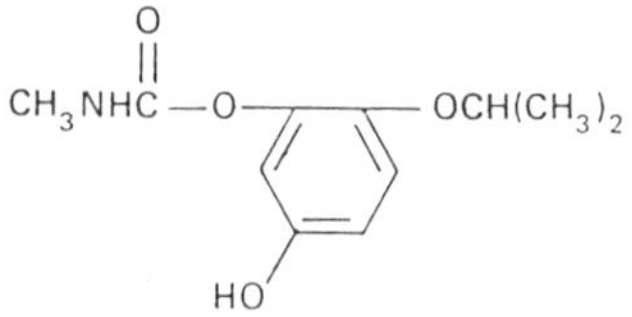

5 - hydroxy propoxur

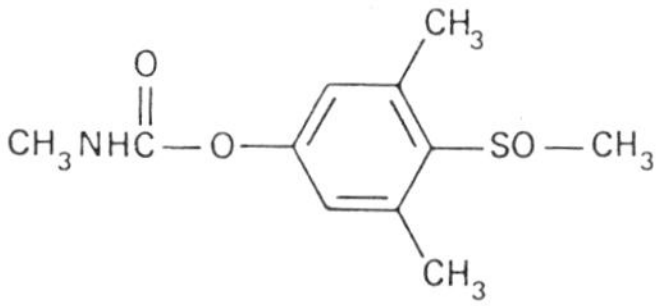

methiocarb sulfoxide

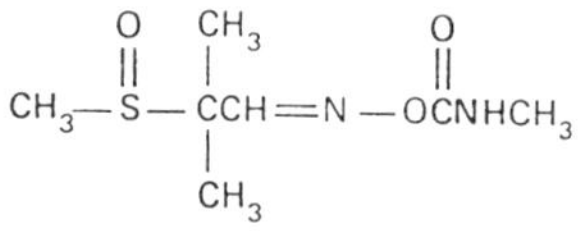

aldicarb sulfoxide

WHO 861207

Fig. 3. Toxic metabolites of methyl carbamate insecticides.

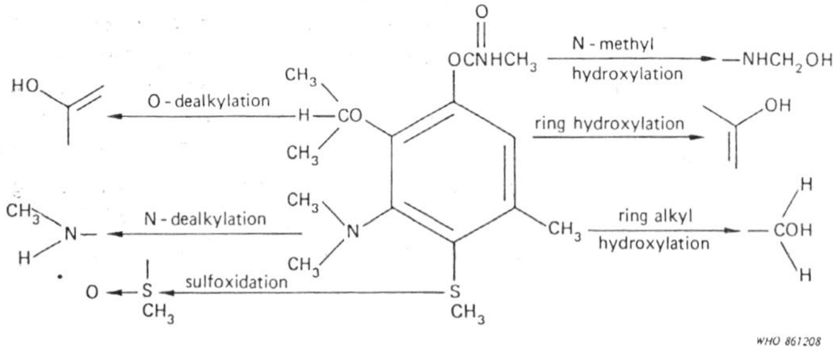

Fig. 4. Sites of oxidation on a hypothetical methyl carbamate insecticide. From: Aldridge & Magos (1978).

Because of the variety of different groups present in carbamate insecticides, the metabolism of these compounds is often complex. Carbaryl, for example, is a relatively simple compound, yet it is metabolized to at least 15 different compounds in mammals through a variety of oxidative and hydrolytic reactions (Leeling & Casida, 1966). A study showed that, as well as the organic solvent-soluble unconjugated metabolites, the urine from carbaryl-treated rabbits contained at least 4 or 5 other water-soluble metabolites. These are probably conjugates of the hydroxylated products of carbaryl, i.e., glucuronides or sulfates.

6.2.3.3 Hydrolysis

Carbamates are hydrolysed either spontaneously or by esterases (Aldridge & Reiner, 1972), yielding, as final products, an amine, carbon dioxide (CO_2), and an alcohol or phenol:

$$R^1HN\text{-}C(O)OR^2 + H_2O \dashrightarrow R^1NH_2 + CO_2 + R^2OH$$

The mechanism of hydrolysis is different for N-methyl and N-dimethyl carbamates.

In general, the rates of hydrolysis of carbamates by esterases are faster in mammals than in plants and insects, though there are exceptions. The differences in enzymatic rates of hydrolysis depend on the structure of the carbamate and on the particular esterase.

Hydrolysis of the carbamates is catalysed by a group of enzymes known as A-esterases or arylesterases (EC 3.1.1.2). It has been established that enzymatic hydrolysis occurs both in vitro and in vivo, but, to date, it has not been determined to what extent the in vivo hydrolysis contributes to the detoxification of the chemical.

The hydrolytic activity of plasma-albumin has also been indicated (Augustinsson & Casida 1959; Casida & Augustinsson, 1959; Reiner & Skrinjaric-Spoljar, 1968).

6.2.3.4 Conjugation

The conversion to conjugated compounds of the hydroxy products produced by the drug-metabolizing enzyme systems is an important reaction that leads to the formation of water-soluble compounds such as O- and N-glucuronides, sulfates, and mercapturic acid, which can be eliminated via the urine or faeces.

6.2.3.5 Examples of biotransformation of carbamates

A few examples of the biotransformation of carbamates are given in Fig. 5.

Aldicarb is unusual in that it does not contain an aromatic ring, and it is also a carbamate ester of an oxime. Knaak et al. (1966) studied its metabolism in rats. The metabolic reactions of aldicarb are similar to the oxidative and hydrolytic pathways found in the thioether-containing organophosphorous esters. The thioether moiety is rapidly oxidized to the oxime-sulfoxide; the oxime-sulfoxide is oxidized much more slowly to the oxime-sulfone. A large number of metabolites resulting from the hydrolysis of the carbamate moiety of the oxidative metabolites of aldicarb have been identified.

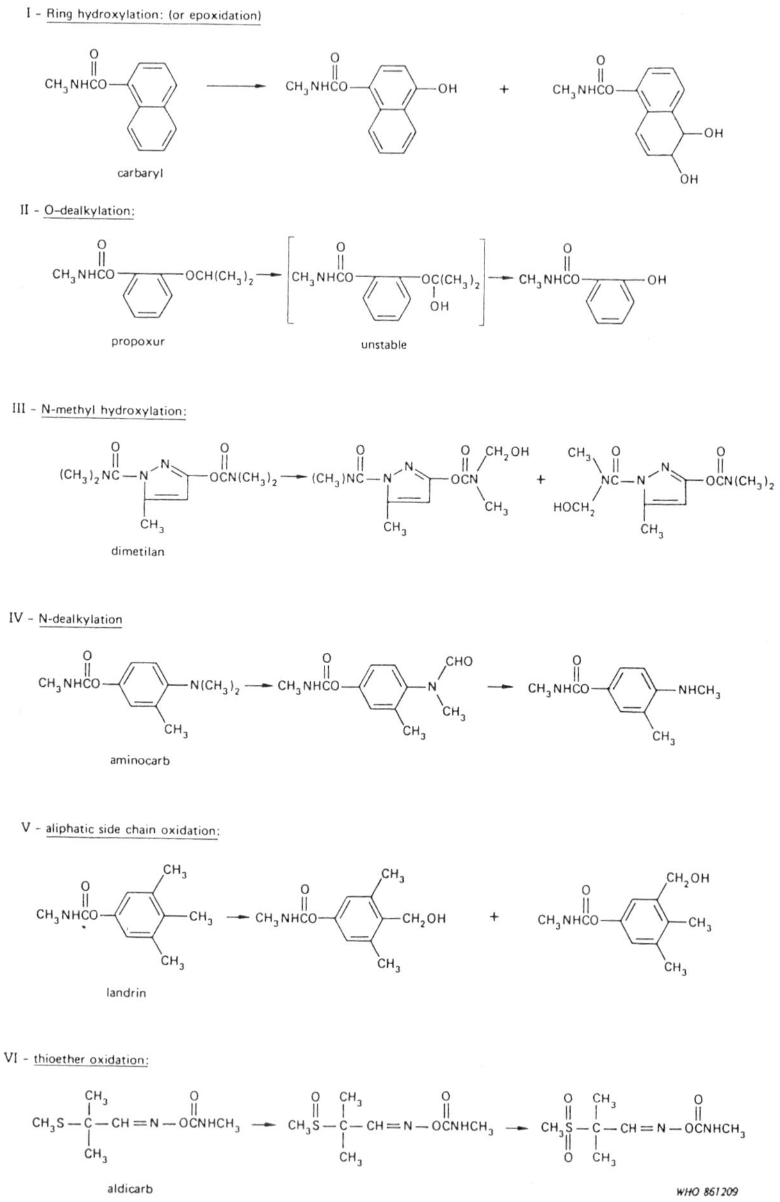

Fig. 5. Carbamate detoxification mechanisms. From: Fukuto (1972).

The oxime-sulfoxide was found to be the major hydrolytic metabolite in plants and insects. Other research workers recovered the nitrile sulfoxide, in greater quantities, both compounds arising from cleavage of the carbamate moiety of aldicarb sulfoxide (Metcalf et al., 1966; Fukuto, 1972).

Propham and chlorpropham have been studied in a number of animal species. Both compounds are rather rapidly metabolized. The major metabolites recovered from the animals were 4-hydroxypropham and 4-hydroxychloropropham, respectively. Both were excreted as the sulfate and/or glucuronide. Depending on the animal species, hydroxylation of propham also occurred in the 2- and 3-position, and even 3,4-dihydroxypropham was identified. Oxidation of the 1-carbon of the isopropyl moiety of chlorpropham resulted in the formation of the monohydroxy product, which was converted to the 1,3-hydroxy and 1-carboxy derivative. Hydrolysis of chloropropham also occurred to a significant extent, since the other major metabolites found were conjugates of 3-chloro-4-hydroxy- and 5-chloro-2-hydroxy-acetanilide. Chlorpropham is more susceptible to hydrolytic degradation than propham, or, conversely, propham is more susceptible to oxidative degradation than chlorpropham (Bobik et al., 1972; Paulson et al., 1973; Fang et al., 1974).

Phenmedipham and desmedipham were hydrolysed mainly into methyl-N or ethyl-N-(3 hydroxyphenyl) carbamates. Both compounds were subsequently hydrolysed to 3-aminophenol. Furthermore, the aminophenol was N-acetylated, and the phenolic hydroxyl moiety was conjugated with the sulfate or glucuronide (Sonawane & Knowles, 1972).

Carbaryl is rapidly hydroxylated or hydrolysed and thereafter conjugated and eliminated from a number of animal species, principally in the urine as glucuronides or sulfates. Depending on the animal species, at least 8 water-soluble metabolites were found. Dorough & Casida (1964) and Fukuto (1972) studied the biotransformation of carbaryl in animals. It was found that hydroxylation took place, and the following metabolites were identified: 1-naphthyl N-hydroxymethylcarbamate (and as a result of ring hydroxylation) 4-hydroxy-1-naphthyl-N-methylcarbamate, 5-hydroxy-1-naphthyl-N-methylcarbamate, and 5,6-dihydroxy-1-naphthylmethylcarbamate.

Certain metabolites appeared to be the hydrolytic products of carbamates with modified ring structures, such as 1-hydroxy-5,6-dihydro-naphthalene and 1-naphthol. The latter compound is found in the urine as 1-naphthyl glucuronide and/or 1-naphthyl sulfate. Sullivan et al. (1972) identified one of these metabolites, 5,6-dihydroxy carbaryl glucuronide, in the urine of rats. Most metabolites of carbaryl are eliminated in a similar pattern in human beings, rats, guinea-pigs, and sheep, but in a different way in dogs (FAO/WHO, 1968b).

The metabolism of benomyl has been studied in the mouse, rat, rabbit, dog, sheep, and cow and is qualitatively the same in all animal species studied.

The basic route of benomyl metabolism involves hydroxylation to 5-hydroxy-2-benzimidazolecarbamate (5-HBC). This is the main metabolite and conjugates (glucuronides or sulfates) of the compound are usually eliminated via the urine, bile, and faeces (Douch, 1973; Gardiner et al., 1974). In contrast, dogs treated orally with ^{14}C-benomyl eliminated only 16% of the radioactivity in the urine and 83% in the faeces. The principal metabolites in the faeces were carbendazim and 5-hydroxycarbendazim. In addition, 4-hydroxycarbendazim was identified as a metabolite in the urine, faeces, and milk of a cow treated with benomyl (Fig. 6).

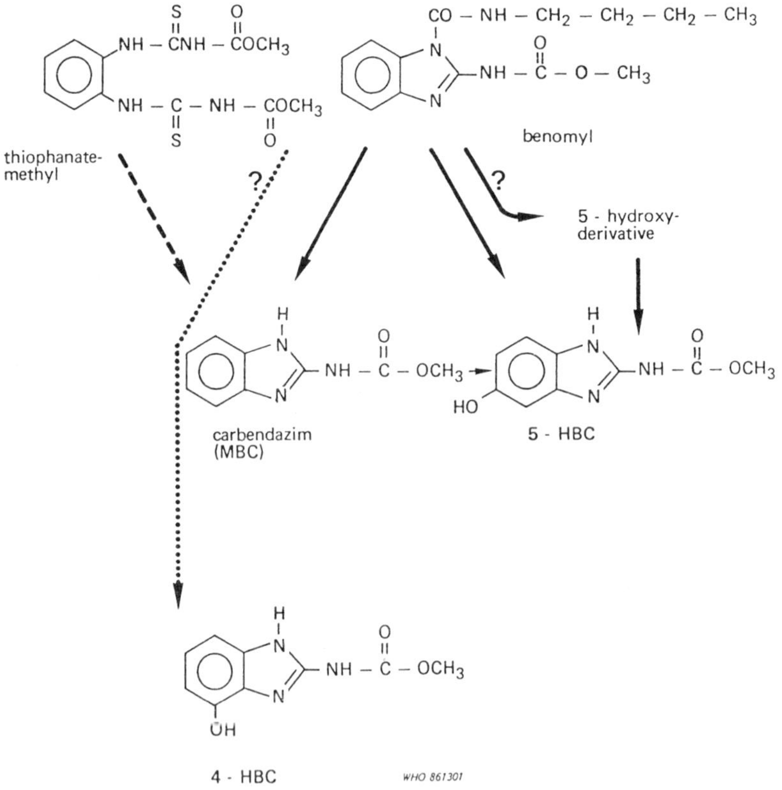

Fig. 6. The major metabolic breakdown products of benomyl, carbendazim, and thiophanate methyl. From: FAO/WHO (1974b, 1978b) and Finnish National Board of Health (1982).

Heybroek et al. (1984) studied the metabolism of [ring-^{14}C] asulam in the rat. Most of the radioactivity (61 - 74%) administered orally or intravenously was excreted in the urine in 24 h as unchanged asulam, 8 - 14% as $\underline{N}^4$-acetylasulam, and up to 2.5% as $\underline{N}^4$-acetylsulfanilamide. Small amounts of radioactivity were found in faeces.

These examples illustrate the different metabolic pathways of biotransformation. For more details, readers should refer to Fukuto (1972) who has reviewed the metabolism of carbaryl, mexacarbamate, aminocarb, formetanate, aldicarb, methiocarb, carbofuran, MPMC, trimethacarb carbanolate, propoxur, and dimetilan. Harvey & Han (1978a) have described the metabolism of oxamyl and Harvey et al. (1973), of methomyl in the rat. The metabolism of pirimicarb was extensively described in FAO/WHO (1977b). Furthermore, information on the metabolism is provided in the handbooks on carbamates, and the different monographs of the Joint FAO/WHO Meetings on Pesticide Residues.

6.3 Elimination and Excretion in Expired Air, Faeces, and Urine

6.3.1 Man

In a study on human volunteers, 2 men were dosed orally with carbaryl in gelatin capsules at 2.0 mg/kg body weight. Chromatographic analyses of a 4-h urine sample revealed 1-naphthyl glucuronide (15%), 1-naphthyl sulfate (8%), and 4-(methylcarbamoyloxy)-1-naphthyl glucuronide (4%). In addition, 1-naphthyl methylimidocarbonate $\underline{O}$-glucuronide was identified by fluorometry (FAO/WHO, 1968b). During the first 8 h, the urinary excretion of these metabolites constituted 12 - 15% of the carbaryl administered. Small amounts of urinary metabolites were detected by fluorometric analysis on the second day, but not afterwards (Knaak et al., 1968). In another study, 1-naphthyl glucuronide, 1-naphthyl sulfate, and other unidentified metabolites were found in 24-h urine samples from men exposed to carbaryl dust during a packaging operation in a factory (details not available) (Knaak et al., 1965).

The determination of 1-naphthol or its sulfate or glucuronide in the urine of persons occupationally exposed to carbaryl is one method of biological monitoring for this exposure (FAO/WHO, 1982b).

Dawson et al. (1964) investigated the metabolism and excretion of propoxur in 3 male volunteers orally administered a dose of 50 mg. Within 8 - 10 h, 27% of the propoxur appeared in the urine as its metabolite 2-isopropoxyphenol, indicating that propoxur is rapidly absorbed and hydrolysed in human beings.

The absorption and fate of benomyl in animals following oral administration of the compound varies with species. According to a study that involved one rat and one dog administered 2-^{14}C-benomyl by stomach tube, the rat eliminated about 86% of a single dose via the urine and 13% via faeces. The dog eliminated only 16% of a similar benomyl dose via the urine, while almost 83% was detected in faeces. No data are available concerning the possible roles of the enterohepatic circulation and biliary excretion of benomyl or its metabolites (Gardiner et al., 1974).

6.3.2 Laboratory animals

In animals, oxidation of carbamates often, but not always, results in detoxification. Oxidative metabolism generally leads to products of greater polarity and water solubility, and these can be more readily eliminated through the urine and faeces than the parent compound.

In the rat, benomyl is mainly excreted via the urine (78.9% of the given dose) as 5-HBC conjugates. In the dog, 99% of the dose administered was eliminated within 72 h. The major route of elimination was via the faeces in which benomyl and/or MBC and 5-HBC were detected. On the basis of studies on the rat, dog, dairy cow, and chicken, neither benomyl nor its metabolites tend to accumulate in animal tissues (Gardiner et al., 1974).

According to FAO/WHO (1974b), feeding of ^{14}C- or ^{35}S-labelled thiophanate-methyl to rats, mice, and dogs resulted in 80 - 100% recovery of the label in the faeces and urine within 96 h. Unmetabolized thiophanate-methyl has been reported to be the major component of faecal elimination. The minor part consisted of 4-hydroxy-thiophanate-methyl, dimethyl-4,4-O-phenylenebisallophanate. MBC and 5-hydroxy-MBC were also detected.

Following application of carbendazim to rats and mice by intragastric intubation, almost all metabolites in the urine were conjugated as sulfate esters. 5-HBC was released as the only metabolite extractable from water. A higher proportion of polar compounds was found in the urine of mice than in that of rats. Polarity was caused by the phenolic hydroxyl group (FAO/WHO, 1985a).

The excretion of ^{14}C-labelled propham and chlorpropham was investigated in female rats after a single oral dose. The average 3-day urinary excretion of radioactivity was between 56 and 85%, depending on the position of labelling: viz chain or ring labelling. With chain ^{14}C-chlorpropham, an average of 35% of the administered radioactivity appeared in the respired air, compared with only 5% for chain ^{14}C-propham (Fang et al., 1974).

6.4 Metabolism in Plants

Studies on the metabolism and fate of carbamate pesticides in plants are of great importance for assessing the human hazards from the consumption of fruit and vegetables containing residues of the applied pesticides and metabolites.

Generally, the metabolism of carbamates in plants follows another route of detoxification. The hydroxylated carbamates produced by enzymes are conjugated either with amino acids (for instance cysteine) or as glycosides and phosphates, which are stored as terminal metabolites (Still & Rusness, 1977; Aldridge & Magos, 1978).

The toxicological properties of plant conjugates in mammals or other animals are not generally known. Insecticidal methyl carbamate esters are vulnerable to hydrolytic cleavage, which results in detoxified products. Detoxification of carbamate esters in plants by hydrolysis occurs to a lesser extent than detoxification by oxidative metabolism; nevertheless, hydrolytic cleavage of the ester moiety is a significant detoxification mechanism. For instance, in sugar beets treated with desmedipham, the major metabolites were ethyl-3-hydroxy phenyl carbamate and 3-aminophenol (Knowles & Sonawane, 1972). Carbetamide is hydrolysed to aniline (Guardigli et al., 1977). Aromatic ring hydroxylation appears to be the predominant metabolic reaction that carbamate herbicides undergo in plants. The major metabolites isolated from soybean plants, after uptake of chlorpropham, were glucoside conjugates of 2-hydroxy- and 4-hydroxypropham. Similar results were obtained with propham (Still & Mansager, 1973a,b, 1975).

There is evidence that the ring hydroxylation varies with the plant species. In other carbamates, for instance in the case of barban, ring hydroxylation did not appear to take place, and polar metabolites were produced, which were hydrolysed into chloroanilines (Still & Mansager, 1972).

Furthermore, oxidation of aliphatic groups takes place in carbamate herbicides (Wiedmann et al., 1976). Still & Herrett (1975) demonstrated <u>N</u>-methyl hydroxylation for dichlormate, a metabolite that is either conjugated or loses the hydroxymethyl moiety to produce <u>N</u>-demethylated dichlormate.

Thiophanate-methyl is mainly metabolized into carbendazim (MBC) and, to a less extent, into 2-aminobenzimidazole (2AB). A few other metabolites have also been found (FAO/WHO, 1974b).

The metabolic pathway for carbaryl in plants is identical, whether the compound is injected into the stem or applied to the leaf surface. After entering the plant, carbaryl undergoes biotransformation to its primary metabolites, which are similar to the ones formed in animals. These hydroxylated metabolites, which are less toxic than carbaryl itself, are conjugated by plants to form water-soluble glycosides. Injection of carbaryl into the bean plant produced water-soluble metabolites attributable to hydroxylation of the ring or N-methyl group followed by conjugation, mainly as glycosides (Kuhr, 1968; FAO/WHO, 1970b; Fukuto, 1972).

The metabolism of aldicarb in plants is the same as that in mammals. In the cotton plant, the thioether moiety is rapidly oxidized to the sulfoxide, and the latter is slowly transformed to the sulfone. The sulfoxide is rather stable and can be present in cotton plants, even as long as 2 months after treatment. The total metabolites present can be as high as 80% (Kuhr, 1968). Because of their persistence in plants, and their even higher anticholinesterase activity compared with aldicarb itself, the sulfoxide and sulfone should be taken into account when considering residue tolerances.

The major metabolite isolated from bean plants, 28 days after treatment with carbofuran, was a conjugate of 3-hydroxy carbofuran, which remained in the aqueous phase after extraction with an organic solvent. The conjugate represented 55% of the total ^{14}C-labelled residues present in the plant. 3-Hydroxy-carbofuran is highly toxic for mammals (LD_{50} in the rat 7 mg/kg body weight). The conjugate may also be toxic, particularly taking into account the acid conditions of the mammalian stomach under which the conjugate will be hydrolysed (FAO/WHO, 1977b, 1980b).

Harvey et al. (1978) studied the metabolism of oxamyl in plants (tobacco, alfalfa, peanuts, potatoes, apples, oranges, and tomatoes). The major route of degradation involved hydrolysis to the corresponding oximino compound, which in turn became conjugated with glucose. Further metabolism resulted in the loss of one of the N'-methyl groups and/or addition of other glucose units to the sugar moiety of the original conjugate. Total breakdown into normal natural products has been demonstrated.

In a study by Harvey & Reiser (1973), methomyl rapidly degraded to carbon dioxide and acetonitrile, which volatilized from the plant tissues. The half-life for methomyl was of the order of 3 - 6 days. The remainder of the compound had been reincorporated into natural plant components.

The metabolic pathways for a number of carbamate pesticides are described in more detail in Fukuto (1972).

7. EFFECTS ON ORGANISMS IN THE ENVIRONMENT

7.1 Microorganisms

As described in section 4.2.1, soil microorganisms are capable of hydrolysing carbamates. Furthermore, it seems that microorganisms can easily adapt themselves to metabolize the different types of carbamates (Aly & El-Dib, 1972). Nevertheless, carbamates and their metabolites can affect the microflora and cause changes that may have an important impact on the maintenance of the soil productivity (Filip, 1974).

7.2 Aquatic Organisms

In general, carbamates are not very stable under aquatic conditions (section 4). The solubility in water differs considerably for the different carbamates (Annex I). Furthermore, photodecomposition takes place and microorganisms are able to degrade these compounds. By these processes, the carbamate is broken down (oxidation, hydrolysis) into other compounds. Some will be less toxic, others even more toxic than the parent compound. Carbaryl will be hydrolysed into 1-naphthol, which is just as toxic (Armstrong & Millemann, 1974a). It seems unlikely that most of the carbamates will persist long in the environment, but there are exceptions, such as dimetilan and carbendazim.

Many data are available on the acute toxicity of the carbamates in several fresh- and salt-water fish and other organisms, some of which are listed in Tables 4 and 5.

When carbamates are deliberately added to an aquatic system, caution is necessary since, with high concentrations, a number of species are at risk. For example, while applications of carbaryl of 0.93 kg/ha were effective in controlling the ghost shrimp (*Callianassa californiensis*), an oyster pest, they also caused a reduction in the population of juvenile clams (Armstrong & Millemann, 1974a).

In a similar study (Armstrong & Millemann, 1974b), carbaryl and its hydrolytic product, 1-naphthol, were examined for other effects on the mussel (*Mytilus edulis*). An age dependence was noted; the most sensitive stage (appearance of the first polar body shortly after fertilization) had EC_{50}s of 5.3 and 5.2 mg/litre for carbaryl and 1-naphthol, respectively. Effects of the compounds on development were characterized by disjunction of blastomeres and asynchronous and unaligned cleavages. Carbaryl, at a single dose of 0.1 mg/litre, also disrupted normal schooling behaviour in juvenile *Menidia medidia*; the disruptive effects were attributed to the 1-naphthol rather than to the parent

Table 4. Acute aquatic toxicity (LC_{50} in mg/litre)[a]

Compound	Type of organism	Name of organism	Stage or weight (g)	Temperature (°C)	Concentration (mg/litre) (after 96 h)
Aldicarb	fish	Rainbow trout (*Salmo gairdneri*)	0.5	12	0.56
	fish	Bluegill (*Lepomis machrochirus*)	1.3	24	0.05
Aminocarb	fish	Rainbow trout (*Salmo gairdneri*)	1.5	12	13.5[b]
	fish	Bluegill (*Lepomis machrochirus*)	2.0	20	3.1[b]
	crustacea	Daphnia (*Daphnia magna*)	first instar	21	0.01 - 0.1[c] (48-h EC_{50})
	fish	Rainbow trout (*Salmo gairdneri*)	1.5	10	0.13[c]
	fish	Bluegill (*Lepomis machrochirus*)	0.6	20	0.1[c]
	insect	Midge (*Chironomus plumosus*)	fourth instar	20	0.27[c]
Benomyl	fish	Rainbow trout (*Salmo gairdneri*)	1.2	12	0.17[d]
	fish	Bluegill (*Lepomis machrochirus*)	0.9	22	0.85[d]
	fish	Rainbow trout (*Salmo gairdneri*)	1.0	12	0.31[e]
	fish	Bluegill (*Lepomis machrochirus*)	0.6	22	1.2[e]
Benomyl metabolite MBC	fish	Rainbow trout (*Salmo gairdneri*)	0.2	12	0.37[f]
Bufencarb	crustacea	Scud (*Gammarus fasciatus*)	mature	15	0.001
	fish	Goldfish (*Carassius auratus*)	1.0	18	0.29
Carbaryl	fish	Rainbow trout (*Salmo gairdneri*)	1.5	12	1.95[d]
	fish	Bluegill (*Lepomis machrochirus*)	1.2	18	6.76[d]
	crustacea	Daphnia (*Daphnia pulex*)	first instar	16	0.0064[d] (48-h EC_{50})
	crustacea	Scud (*Gammarus fasciatus*)	mature	21	0.026[d]

Table 4 (contd).

Carbofuran	fish	Rainbow trout (*Salmo gairdneri*)	1.5	12	0.38[d]
	fish	Bluegill (*Lepomis machrochirus*)	0.8	18	0.24[e]
Dichlormate	fish	Rainbow trout (*Salmo gairdneri*)	0.8	12	4.9
Methiocarb	fish	Rainbow trout (*Salmo gairdneri*)	1.3	12	0.80
	fish	Bluegill (*Lepomis machrochirus*)	1.0	24	0.21
Methomyl	fish	Rainbow trout (*Salmo gairdneri*)	1.1	12	1.60[b]
	fish	Bluegill (*Lepomis machrochirus*)	0.9	20	1.05[b]
	crustacea	Daphnia (*Daphina magna*)	first instar	21	0.009[b] (48-h EC_{50})
Mexacarbate	fish	Rainbow trout (*Salmo gairdneri*)	1.0	11	12.0[g]
	fish	Bluegill (*Lepomis machrochirus*)	0.7	12	22.9[g]
	crustacea	Daphina (*Daphina pulex*)	first instar	15	0.010[g] (48-h EC_{50})
	crustacea	Scud (*Gammarus fasciatus*)	mature	-	0.04[g]
Trimethacarb	fish	Rainbow trout (*Salmo gairdneri*)	1.2	12	1.0
	fish	Bluegill (*Lepomis machrochirus*)	0.9	18	11.6

a From: Johnson & Finley (1980).
b Technical material, 95 - 98%.
c Liquid formulation, 17%.
d Technical material, 99%.
e Wettable powder, 50%.
f Methyl-2-benzimidazole (MBC), 99%.
g Technical material, 90 - 95%.

Note: It should be recognized that the LC_{50} and EC_{50} values only give an indication of the toxicity and that the toxicity for these organisms may be appreciably changed by variations in temperature, pH, oxygen content, and water hardness. A 10-fold increase in the toxicity may be found. Also, the life stage of the organisms is an important factor in estimating the toxicity of the compound.

Table 5. Acute toxicity of carbamate pesticides for some aquatic organisms

Pesticide[a]	TLMs for organisms at indicated time (mg/litre)					Reference
	Carp (*Cyprinus carpio* Linné)	Goldifsh (*Carassius auratus*)	Killifish (*Fundulus* sp)	Guppy (*Lebistes reticulatus* Peters)	Water flea (*Daphnia pulex* Leydig)	
	48 h	48 h	48 h	48 h	3 h	
Benomyl	7.5	12	11 (wettable powder)	-	14	Yoshida & Nishiuchi (1972)
BPMC	16	10 ~ 40	1.7	5.0	0.32	Yoshida & Nishiuchi (1972); Nishiuchi (1974)
Carbaryl	13	10 ~ 40	2.8	2.8	0.05	Yoshida & Nishiuchi (1972)
Carbendazim	> 40	-	40	-	> 40	Yoshida & Nishiuchi (1976)
CPMC	10 ~ 40	10 ~ 40	5.6 (emulsifiable concentrate)	7.0	0.1 ~ 0.5	Yoshida & Nishiuchi (1972); Nishiuchi (1974)
Isoprocarb	10 ~ 40	10 ~ 40	5.9	3.0	0.30	Yoshida & Nishiuchi (1972); Nishiuchi (1974)
Mecarbam	0.70	0.68	0.35	0.055 granule	0.03	Yoshida & Nishiuchi (1972); Nishiuchi (1974)
Methomyl	2.8	2.7	0.87	1.0	0.045	Yoshida & Nishiuchi (1972); Nishiuchi (1974)

Table 5 (contd).

MPMC	10 ~ 40	10 ~ 40	12	7.3	0.07	Yoshida & Nishiuchi (1972); Nishiuchi (1974)
MTMC	10 ~ 40	10 ~ 40	27	13	0.35	Yoshida & Nishiuchi (1972); Nishiuchi (1974)
Pirimicarb	> 40	-	40	-	0.048	Yoshida & Nishiuchi (1976)
Promecarb	2.7	-	3.3	-	0.020	Yoshida & Nishiuchi (1976)
Terbam	0.92	-	2.2	-	0.025	Yoshida & Nishiuchi (1976)
Thiophanate	> 40	> 40	> 40	-	> 40	Yoshida & Nishiuchi (1972)
Thiophanate-methyl	11	> 40	11 (wettable powder)	-	> 40	Yoshida & Nishiuchi (1972)
XMC	> 40	> 40	33	25	0.055	Yoshida & Nishiuchi (1972); Nishiuchi (1974)

a BPMC = 1-sec-butylphenylmethyl carbamate
CPMC = 1-chlorophenylmethyl carbamate
MPMC = 3,4-xylylmethyl carbamate
MTMC = 4-tolylmethyl carbamate
Terbam = 4-tert-butylphenylmethyl carbamate
XMC = 3,5-xylylmethyl carbamate

Note: Test methods are officially recognized methods based on the Notification of the Ministry of Agriculture, Forestry, and Fisheries of Japan.

carbamate. These effects were reversed within 3 days of placing the fish in clean water (Weis & Weis, 1974). Hansen (1969) studied the capacity of sheepshead minnows (Cyprinodon variegatus) to avoid carbaryl. The fish did not avoid carbaryl in concentrations of 0.1 - 10 mg/litre water.

Carbamates can have adverse effects on algae. Stadnyk et al. (1971) reported that carbaryl at a concentration of 0.1 mg/litre caused an increase in cell numbers and in the biomass of the green algae Scenedesmus quadricaudata.

Motsuage fish (Motsugo pseudoras bora parva) were reported to accumulate carbaryl and BPMC during a 30-day exposure to a concentration of 0.6 - 1.2 mg/litre. Carbaryl uptake was greatest. However, this compound underwent more rapid metabolism and excretion than BPMC. Metabolism of BPMC was slow and introduced permanent spinal curvature of the backbone in about 30% of the fish exposed, an effect seen with the organophosphorus insecticide diazinon (Kanazawa, 1975).

A residue study indicated that over a 28-day period at the highest exposure level of 0.75 mg methomyl/litre, no accumulation of the carbamate took place. During the entire exposure period, the values ranged from 0.36 to 0.45 mg/kg. After a 3-day withdrawal period, no methomyl (< 0.02 mg/kg) was detected in the fish (Kaplan & Sherman, 1977).

The results summarized in Table 4 are representative of the toxicity of the different carbamates for fish and invertebrate species. In general, the LC_{50}s of carbamates for different fish species range from approximately 0.1 to 10 mg/litre. Invertebrates are usually more sensitive. The EC_{50} values for Daphnids and other species are mainly below 0.1 mg/litre. Some carbamates seem to be very toxic (< 0.01 mg/litre) for these invertebrates.

7.2.1 Field studies

Little information is available concerning the fate and persistence of carbamates in the aquatic environment. The presence of these carbamates in surface waters may have an effect on water organisms.

Quraishi (1972) studied the persistence of aldicarb. Field water treated in the laboratory at 100 mg aldicarb/litre resulted in residues of aldicarb and its metabolites of 0.4 mg/litre after 11 months. Water was stored at 16 - 20 °C and exposed for approximately 507 h to sunlight.

Data concerning the fate of aminocarb have been reviewed by Vassilieff & Ecobichon (1982). A half-life of 28.5 days in pond water was found under normal environmental conditions. The principal metabolite appeared to be the phenol, though the methylamino and formylamino analogues were also observed.

Harvey & Pease (1973) found that, under field conditions in Delaware, Florida, and North Carolina, methomyl broke down almost completely within 1 month; most of it was lost from the soil by volatilization, presumably as carbon dioxide. Small amounts extracted from the soil consisted of methomyl, S-methyl N-hydroxythioacetimidate, and some polar compounds. A run-off study, under farm conditions, showed that the compound did not move into untreated areas with run-off water.

7.3 Terrestrial Organisms

7.3.1 Effects on soil fauna

Thiophanate-methyl and carbendazim have been shown to be equally as toxic through contact as benomyl. Earthworms (*Lumbricus terrestris*) immersed for 1 min in a 0.6% aqueous suspension of these compounds died in 14 days. Worms in pots containing soil drenched at a rate of 0.78 g/m^2 died within 18 days (Wright & Stringer, 1973).

7.3.2 Wildlife

The LD_{50} of benomyl for mallard ducks and quail is > 5000 mg/kg body weight.The cumulative toxicity of formetanate (7 days) is approximately 6800 mg/kg body weight for ducks, and > 4640 mg/kg body weight for quail and for pheasant (Aldridge & Magos, 1978).

Lethal and sub-lethal doses of aldicarb, methiocarb, oxamyl, pirimicarb, and thiofanox administered to Japanese quail produced significant inhibition of plasma- and brain-ChE and influenced the activity of other enzymes, such as alpha-naphthyl acetate esterase, glutamate dehydrogenase, and glutamate oxaloacetic transaminase. With sub-lethal dose levels, plasma-ChE activity recovered within a few days (Westlake et al., 1981).

An 8-day dietary LC_{50} for the Peking duck was 1890 mg/kg feed; for bobwhite quail, the dietary LC_{50} was 3680 mg/kg feed (Kaplan & Sherman, 1977).

Bobwhites (*Colinus virginianus*) were provided diets containing sublethal levels of carbaryl (237 or 1235 mg/kg), for 7 days, or carbofuran (26 mg/kg), for 14 days. Food intake, body weight, and the locomotor activity of adult bobwhites were normal. Administration of a diet containing 131 mg carbofuran/kg resulted in reductions in food intake, body weight, and locomotor activity (Robel et al., 1982).

7.3.3 Bees

An undesirable effect caused by the use of carbamate insecticides has been the mortality of honey bees, a species that exhibits a high sensitivity to these compounds. Because of the agricultural and ecological importance of bees, thorough comparative toxicity studies have been carried out on a large number of compounds to identify those with the widest margin of safety (Atkins et al., 1973).

Abdel-Aal & Fahmy (1977) compared the effects of aldicarb, methomyl, N-desmethylmethomyl, carbaryl, and carbofuran on the honey bee (Apis mellifera) and found that all of the compounds were extremely toxic.

7.4 Earthworm and Mite Populations

Studies by Stringer & Wright (1973) and Wright & Stringer (1973) showed that earthworm populations in apple orchard plots sprayed with either benomyl or thiophanate-methyl were reduced. In particular, Lumbricus terrestris, an important species in the ecology of the apple orchard, was virtually eliminated. Captive worms would not feed on leaf material that had been sprayed at 1.75 $\mu g/cm^2$ with benomyl, methyl benzimidazol-2-yl carbamate (MBC), or thiophanate-methyl, and feeding was significantly reduced at a level of 0.87 $\mu g/cm^2$. All the worms were killed following contact with benomyl as an aqueous suspension or soil drench (7.75 kg/ha).

It was suggested by the authors that the toxicity of benomyl might be due to the anticholinesterase activity of the carbamate moiety. Effects such as lethargic movements, muscular paralysis, and death support this supposition.

The spraying of benomyl and thiophanate-methyl in orchards reduced the number and biomass of all earthworm species combined and also of each individual species (Stringer & Lyons, 1974). It was observed that over 90% reduction in earthworm populations occurred in pastures treated with benomyl. It was suggested that these changes were reversible, and that the populations of earthworms would return to normal a few years after the termination of the treatment (Tomlin & Gore, 1974).

The carbamate insecticide most commonly used in the soil is carbaryl; it is used mainly in woodlands and orchards where the formation and fertility of the soil are important.

Stegeman & LeRoy (1964) applied carbaryl at 1.3, 11.2, and 56 kg ai/ha to a red pine plantation and mixed hardwood stands, and, though the lowest dose did not have any effects on mite and Collembola populations, the other 2 dose levels considerably decreased the numbers of mites and Collembolla. Populations began to recover after 2 months, but Collembola were more sensitive and recovered more slowly. There seems to be little likelihood of long-term effects of carbaryl on mite populations (Edwards & Thompson, 1973).

8. EFFECTS ON EXPERIMENTAL ANIMALS AND *IN VITRO* TEST SYSTEMS

8.1 Single Exposures

Short- and long-term toxicity studies with carbamates have been carried out for a period of 20 - 30 years. It is clear that the production of these carbamates has improved during this period and that, consequently, purer products have become available. The use of less pure substances in earlier toxicity studies could explain the conflicting results between these and later studies.

8.1.1. Oral

Acute oral and dermal toxicity data on animals, for most of the carbamate pesticides, are given in Table 6. The WHO Recommended Classification of Pesticides by Hazard is also cited. This classification is based primarily on the acute oral and dermal toxicity of the technical material for the rat (WHO, 1984). The oral LD_{50}s range from less than 1 mg/kg body weight for the highly-toxic aldicarb to over 5000 mg/kg body weight for the non-insecticidal carbamates such as phenmedipham, carbendazim, propham, and benomyl.

A dose-effect relationship is generally noticed between the administered dose, the severity of effects, and the amount of ChE inhibition. A correlation also exists between the duration of symptoms and the *in vivo* persistence of the compound.

In general, the toxicological effects produced by carbaryl are typical of those produced by methyl carbamates. The effects of carbaryl on a number of different animal species were studied by Carpenter et al. (1961). Rats given a single oral dose of 560 mg/kg body weight showed 42% and 30% inhibition of erythrocyte- and brain-ChE, respectively, within 1/2 h of treatment; only about 5% inhibition of plasma-ChE was observed. All ChE levels were almost normal after 24 h. In studies on rats administered single oral doses ranging from 5 to 50 mg/kg body weight, a decrease in ChE activity occurred within 5 min. Recovery was quite rapid (Krechniak & Foss, 1982).

In 4 dogs, marginal inhibition of erythrocyte-ChE activity was seen after treatment with a single oral dose of 375 mg carbaryl/kg body weight. After 30 min, signs of poisoning were observed including salivation, lachrymation, constriction of pupils, urination, defaecation, increase in respiratory rate, muscular twitching, tremors, and mild convulsions. The animals appeared to have completely recovered the following

Table 6. Acute oral and dermal toxicity data for a number of carbamate pesticides

Carbamate	LD_{50} (mg/kg body weight)[a] oral	dermal	WHO Recommended Classification of Pesticides by Hazard[b]
aldicarb	0.9	> 10.0 (rabbit)	IA
aldoxycarb	26.8	700 - 1400	-
allyxycarb	90 - 99	500	-
aminocarb	30 - 40	275	IB
asulam	> 4000	> 1200	-
barban	1376 - 1429	> 1600 > 20 000 (rabbit)	-
BPMC	623 - 657	> 5000	-
bendiocarb	40 - 156	566 - 600	II
benomyl	> 10 000 > 1000 (dog)	> 10 000 (rabbit)	0
bufencarb	87	680 (rabbit)	-
butacarb	NA	NA	-
carbanolate	NA	NA	-
carbaryl	approximately 500 - 600	> 4000 > 2000 (rabbit)	II
carbendazim	> 15 000 > 2500 (dog) > 10 000 (quail)	> 2000 > 10 000 (rabbit)	0
carbetamide	10 000 1250 (mouse) 1000 (dog)	> 2000 > 500 (rabbit)	-
carbofuran	6 - 14 15 - 19 (dog)	3400[c] (rabbit)	IB
chlorbufam	2500	NA	-
chlorpropham	5000 - 8000 5000 (rabbit)	10 200[d] 2000 (dog)	0
cloethocarb	35.4	4000	-
desmedipham	> 10 250	2000 - 10 000 2025[e] (rabbit)	-

Table 6 (contd).

Carbamate	LD_{50} (mg/kg body weight)[a]		WHO Recommended Classification of Pesticides by Hazard[b]
	oral	dermal	
dimetilan	64 60 - 65 (mouse)	> 2000	-
dichlormate	NA	NA	
dioxacarb	72	approximately 3000 1950 (rabbit)	-
ethiofencarb	411 - 499 224 - 256 (mouse) 155 (quail)	> 1150	II
formetanate	21, 18 (mouse) 19 (dog)	> 5600 > 10 200 (rabbit)	
hoppicide	NA	NA	-
isoprocarb	403 - 485 487 - 512 (mouse) ca 500 (rabbit)	> 500	
karbutilate	3000	> 15 400	-
methiocarb	100 40 (guinea-pig)	350 - 400	II
methomyl	17 - 24 28 (hen)	> 5000 (rabbit)	IB
metolcarb	498 - 580	> 2000	-
mexacarbate	15 - 63	> 500	-
MPMC	380	> 1500 (mouse)	-
MTMC	268 (mouse)	6000	-
oxamyl	5.4 4.2 (quail)	710 (rabbit)	IB
phenmedipham	> 8000 > 4000 (dog) > 3000 (chicken)	> 4000	-
pirimicarb	147 (101 - 210) 107 (mouse) 100 - 200 (dog) 25 - 50 (poultry)	> 500 (rabbit)	II
promacyl	1220 2000 - 4000 (mouse) 4000 - 8000 (hen)	> 4000	-

Table 6 (contd).

Carbamate	LD_{50} (mg/kg body weight)[a] oral	dermal	WHO Recommended Classification of Pesticides by Hazard[b]
promecarb	61 - 90	> 1000 > 1000 (rabbit)	-
propham	9000 3000 (mouse)	6800[f] (rabbit)	O
propoxur	80 - 191 100 - 109 (mouse) 40 (guinea-pig)	1000 - 2400	II
swep	NA	NA	-
terbucarb	NA	NA	-
thiodicarb	NA	NA	-
thiofanox	8.5 43 (quail)	39 (rabbit)	IB
thiophanate-ethyl	> 15 000	> 15 000	-
thiophanate-methyl	6640 - 7500	> 10 000	O
trimethacarb	NA	NA	-
xylylcarb	325 - 375	> 1000	-
XMC	542 445 (rabbit)	-	-

a Data given for rats, unless otherwise stated.

b From: WHO (1984). See this reference for classification of organophosphates not mentioned in this annex.
The hazard referred to in this Classification is the acute risk for health (that is, the risk of single or multiple exposures over a relatively short period of time) that might be encountered accidentally by any person handling the product in accordance with the directions for handling by the manufacturer or in accordance with the rules laid down for storage and transportation by competent international bodies.
Classification relates to the technical material, and not to the formulated product:
Class IA = Extremely hazardous.
Class IB = Highly hazardous.
Class II = Moderately hazardous.
Class III = Slightly hazardous.
Class O = unlikely to present acute hazard in normal use.

c With 75% water-dispersible powder.

d With 50% emulsifiable concentrate formulation.

e With 16% formulation.

f With 25% emulsifiable concentrate formulation.

NA = Not available.

Note: References are the reports of the different WHO/FAO Joint meetings on pesticides residues (Annex III), Kaplan & Sherman (1977), and Worthing & Walker (1983).

day. These signs of poisoning are classic of overstimulation of the parasympathetic nervous system (Carpenter et al., 1961).

Vassilieff & Ecobichon (1983) studied the influence of a single dose of 25 mg aminocarb/kg body weight on the activity of erythrocyte- and brain-AChE, plasma-ChE, and hepatic carboxylesterases in rats. Significant inhibition of all the esterases was observed, 30 min or more after the administration of aminocarb. This severe, but transient, inhibition of the tissue esterases had recovered after 24 h.

8.1.2 Dermal

Acute dermal toxicity values for most carbamates are mainly above 500 mg/kg body weight with the exception of, for instance, aldicarb, which is highly toxic.

8.1.3 Inhalation

Exposure to carbaryl (50% wettable powder, average particle size, 15 µm) at a mean concentration of 390 mg/m^3 (range 344 - 722 mg/m^3), for 4 h, produced nasal and ocular irritation in 6 guinea-pigs. When another group of guinea-pigs inhaled a mean of 230 mg carbaryl/m^3 (average particle size, 5 µm), for 4 h, no clear effects were observed. Single exposures of dogs ranging from 2 to 5 h to a concentration of 75 mg carbaryl/m^3 (average particle size, 5 µm) caused clinical signs of ChE inhibition (Carpenter et al., 1961). Cats exposed to 80 mg carbaryl/m^3 for 6 h showed salivation, excitation, respiratory difficulties, and 39 - 55% inhibition of ChE activity in the serum and 53 - 71% in the erythrocytes. A single exposure to a level of 20 mg/m^3 for 6 h did not induce signs of intoxication, but serum- and erythrocyte-ChE activities were slightly inhibited (Yakim, 1967).

8.2 Short- and Long-Term Exposures

A number of short-term studies have been carried out with different carbamates and a variety of laboratory animal species. In the context of this introduction, it is not possible to mention all the reports available. Details on the short- and long-term toxicities of the individual carbamates will be provided in future Environmental Health Criteria. Most carbamates have been discussed by the Joint FAO/WHO Meetings on Pesticide Residues and monographs have been published (see reference list). Furthermore, the International Agency for Research on Cancer has published a monograph in which a number of carbamates have been evaluated (IARC, 1976).

In the context of this introduction, a number of examples are given to illustrate dose-response relationships and/or target organs or tissues for a number of carbamates.

8.2.1 Oral

Short-term tests on rats or dogs exposed to repeated oral doses of carbamate insecticides or fungicides (such as carbaryl, propoxur, formetanate, carbofuran, and benomyl), revealed inhibition of plasma (serum)-, erythrocyte-, and brain-ChE (Carpenter et al., 1961; Ivanova-Chemishanska & Antov, 1976; Krechniak & Foss, 1982).

(a) Pirimcarb

In a study on 4 dogs (2 males and 2 females) treated with 25 or 50 mg pirimicarb/kg body weight per day, for 110 weeks, 2 of the animals developed anaemia. Haemoglobin content was decreased, methaemoglobin was present, and the reticulocyte percentage was significantly elevated. Levels of unconjugated bilirubin in serum were increased. The 2 other animals were unaffected. There was some indication that pirimicarb was able to induce the production of immune IgG antibodies (Jackson et al., 1977).

(b) Carbaryl

Kidney alterations, such as cytoplasmic vacuolization, were observed in the proximal tubules in short-term studies on monkeys and rats adminstered carbaryl at dose levels of up to 600 or 300 mg/kg body weight, respectively (Coulston, 1967; cited in FAO/WHO, 1970b).

Two pigs from one litter were fed a ration containing carbaryl at the rate of 150 mg/kg body weight for 72 and 83 days, respectively. Three pigs in the same litter were fed 150 mg/kg body weight for 4 weeks and, thereafter, 300 mg/kg body weight, the total feeding period being 46 or 85 days. Two other pigs were maintained on the basal ration, as controls. The clinical syndrome of chronic intoxication was characterized by progressive myasthenia, incoordination, ataxia, tremors, and chronic muscular contractions terminating in paralysis and death. The lesions were confined to the central nervous system and skeletal muscle. Moderate to severe oedema of the myelinated tracts of the cerebellum, brain stem, and upper spinal cord was associated with vascular degenerative changes, consistent with a vasogenic oedema (Smalley et al., 1969). Spontaneous recovery from the effects of sub-lethal doses occurred in about 8 h, indicating the ready reversibility of the carbaryl-enzyme reaction (Carpenter et al., 1961).

Carbaryl was administered to dogs at dose levels of 0, 0.45, 1.8, or 7.2 mg/kg body weight per day, for 5 days/week, for one year. Cloudy swelling of the proximal convoluted and loop tubules of the kidneys was found in the group given 7.2 mg/kg body weight. There were no abnormalities in the composition of the blood, and no ChE inhibition was found. It should be noted that the dose levels tested were low (Carpenter et al., 1961).

In a long-term (2-year) toxicity study on CF-N rats, levels of 0, 50, 100, 200, or 400 mg carbaryl/kg diet (equivalent to 0, 2.5, 5, 10, or 20 mg/kg body weight) were tested. The highest dose level produced cloudy swelling in the tubules of the kidneys (Carpenter et al., 1961).

(c) Carbendazim

Groups of ChR-CD male rats (6/dose) were intubated with 200, 3400, or 5000 mg carbendazim/kg body weight per day, 5 times per week, for 2 weeks. In all groups, the absolute and relative weights of the kidneys were lower, and adverse effects on the testes and reduction or absence of sperm in the epididymides were found (FAO/WHO, 1985a).

(d) Benomyl

In a 2-year toxicity study, Charles River rats were administered dietary levels of benomyl of 0, 100, 500, or 2500 mg/kg (equivalent to 0, 5, 25, or 125 mg/kg body weight). No increase in mortality and no other compound-related effects on haematology, urinalysis, organ weight, or histology were found (Haskell Laboratories, 1969 cited in FAO/WHO, 1985a).

In another 2-year study, dogs were administered diets containing 0, 100, 500, or 2500 mg benomyl/kg (equivalent to 0, 2.5, 12.5, and 62.5 mg/kg body weight). Liver cirrhosis, bile duct proliferation, and testicular degeneration were found at the highest dose level; however, 100 and 500 mg/kg diet did not show any clear effects (Haskell Laboratories, 1970 cited in FAO/WHO, 1985a).

(e) Thiophanate-methyl

An oral toxicity study in which thiophanate-methyl at dose levels of 0, 2, 10, 50, or 250 mg/kg body weight (in capsules) was administered to beagle dogs (8 or 10 animals per group) was conducted for 24 months. Retardation of growth was seen in both sexes in the 250 mg/kg group. Both male and female dogs in the 50 and 250 mg groups showed a significant and dose-related increase in the weight of the thyroid. However, there were no changes in the serum-protein-bound iodine (PBI) levels and no histological differences in the thyroid were found between the control group and the 2 highest dose groups. No other abnormalities were observed (Hashimoto & Fukuda, 1972 cited in FAO/WHO, 1974b).

(f) Methomyl

Groups of rats were given diets containing 0, 10, 50, 125 (increased to 500), or 250 mg methomyl/kg for 90 days. Weight gain in animals administered 250 and 500 mg/kg was slightly suppressed. No clear changes in the different parameters including plasma- and erythrocyte-ChE activity were found (Kaplan & Sherman, 1977).

The same authors carried out studies over 90 days and 2 years on beagle dogs. Dietary levels of 0, 50, 100, and 400 mg methomyl/kg did not result in any nutritional, clinical, haematological, urinary, biochemical, or pathological changes. In the 2-year feeding study, histopathological alterations were seen in kidneys and spleen of animals receiving 400 and 1000 mg methomyl/kg diet. Histopathological changes were seen in the liver and bone marrow in animals administered 1000 mg/kg diet (Kaplan & Sherman, 1977).

The above-mentioned studies give some indication that carbamates may have different mechanisms of action on target organs that are related to substituents other than the carbamate ester moiety (such as the imidazole moiety).

Apart from the effects on ChE activity, effects have been found on the liver, kidneys, thyroid, and testes and also on the activities of other enzyme systems such as ATPase, Gl-6-ase, LDH, GPT, Alk-Pase, etc. (Haskell Laboratories, 1968 cited in FAO/WHO, 1985a; Ivanova-Chemishankska & Antov, 1976).

8.2.1.1 Further information on short- and long-term toxicity

It is not the purpose of this introductory document to provide all the available data on long-term toxicity. However, these long-term studies are important for the establishment of the no-observed-adverse-effect level and consequently the Acceptable Daily Intake (ADI) for man. Thus, it is necessary to indicate the studies that were considered adequate and used by the different FAO/WHO Joint Meetings on Pesticide Residues. The International Agency for Research on Cancer has also evaluated long-term/carcinogenicity studies of a number of carbamates. The results of these international evaluations are summarized in Annex II and III.

From these Annexes, it is clear that aldicarb is the most toxic of the carbamates considered; the established ADI for this compound is 0.001 mg/kg body weight.

The ADIs for the other carbamates may vary from 0.01 to 0.1 mg/kg body weight. These quantities are equivalent to a daily intake for man (60 kg body weight) of approximately 0.6 - 6 mg.

8.2.2 Dermal

(a) Pirimicarb

Daily dermal applications of 500 mg pirimicarb/kg body weight to the skin of rabbits for 24 h over a 14-day period did not produce any signs of intoxication (FAO/WHO, 1977b).

(b) Methomyl

Methomyl applied to the intact skin of rabbits at a repeated rate of 200 mg/kg per day caused mild signs of intoxication such as nasal discharge, wheezing, and transient diarrhoea. All animals survived 15 applications. However, when applied on the abraided skin, much more serious systemic signs were seen, such as nasal discharge, salivation, tremors, poor coordination, and abdominal hypertonia; a few animals died (Kaplan & Sherman, 1977).

(c) Carbaryl

Application of carbaryl at 500 mg/kg body weight to the skin of 6 rabbits resulted in 39% inhibition of serum-ChE activity and 36% inhibition of red cell-ChE activity during the first 24 h following treatment. ChE activity returned to initial levels in 72 h (Yakim, 1967).

8.3 Skin and Eye Irritation; Sensitization

8.3.1 Skin irritation

Rabbits exposed to a single dermal application of carbaryl at a concentration of 100 g/litre in acetone did not show any skin irritation (Carpenter et al., 1961).

8.3.2 Eye irritation

Undiluted carbaryl or solutions of different concentrations were applied to the eyes of rabbits. Technical carbaryl applied in excess as a 100 g/litre suspension in propylene glycol produced mild injury. A 25% aqueous suspension of a microfine material did not cause any injuries; 50 mg of the dust caused only traces of corneal necrosis (Carpenter et al., 1961).

8.3.3 Skin sensitization

(a) Carbaryl

Groups of albino guinea-pigs were given 8 intracutaneous injections (3 per week) of 0.1 ml of a 0.1% dispersion of carbaryl in 3.3% propylene glycol. After a 3-week incubation period, followed by a challenge dose, no clear sensitization reaction was noticed (Carpenter et al., 1961).

(b) Benomyl and thiophanate-methyl

It has been shown that benomyl and thiophanate-methyl can cause slight reversible skin irritation and weak sensitization in guinea-pigs (FAO/WHO, 1974, 1985a).

8.4 Inhalation

(a) Carbaryl

The inhalation toxicity of carbamates for mammals has not been widely studied.

The effects of acute inhalation exposure to carbaryl dust have been examined in rats (10 mg/m^3), guinea-pigs (390 mg/m^3), and dogs (75 mg/m^3) (Carpenter et al., 1961) (section 8.1.3). No treatment-associated, gross or microscopic lesions were seen in guinea-pigs and dogs, even though the exposure was sufficient to cause ChE inhibition. Haemorrhagic areas in the lung were seen in guinea-pigs at the highest level tested (390 mg/m^3).

Four cats exposed to carbaryl in air at 63 mg/m^3 for 6 h per day for 1 month developed signs of cholinergic stimulation during the first 2 h of exposure each day. Periodic salivation was noted. One cat died with marked signs of poisoning on day 20. There was already a significant decrease in serum- and erythrocyte-ChE activity after the first exposure. Exposure to 40 mg/m^3 for 6 h per day for 2 months was reported to impair conditioned reflexes in cats. On certain days, the ChE activity dropped to below 50% of pre-exposure values. Exposure to concentrations of 16 mg/m^3, for 6 h per day, for 4 months did not result in any signs of toxicity (Yakim, 1967). Carpenter et al. (1961) did not find any increase in mortality or "grossly-visible injury" when rats were exposed to a mean concentration of 10 mg/m^3 (range 5 - 20 mg/m^3), for 7 h per day, 5 days/week, over 90 days.

(b) Propoxur

Male and female rats were exposed to propoxur aerosols at average concentrations of 0, 5.7, 10.7, or 31.7 mg/m^3, for 6 h daily, 5 days/week, over a period of 12 weeks. Depression of plasma-, erythrocyte-, and brain-ChE activity was observed at the highest dose level (Kimmerle & Iyatomi, 1976).

8.5 Reproduction, Embryotoxicity, and Teratogenicity

A considerable number of reproduction and teratogenicity studies have been carried out on different carbamates in several animal species. A number of these will be used to show the state of the art giving the well-known carbamates, such as carbaryl, benomyl, carbendazim, and thiophanate-methyl, as examples.

For more details concerning different carbamates, the reader is referred to the reports of the Joint FAO/WHO Meetings on Pesticide Residues and, in the future, to the Environmental Health Criteria Documents on the individual carbamates.

8.5.1 Reproduction

(a) Methomyl

A 3-generation reproduction study with methomyl was carried out on rats. Dose levels of 0, 50, or 100 mg/kg diet were administered for 3 months. There was no evidence from the different indices of reproduction, lactation, and weanling body weights to suggest that methomyl affected reproduction or lactation performance in the litters of three generations (Kaplan & Sherman, 1977).

(b) Benomyl

The effects of benomyl on the gonads and reproductive function were studied in a 3-generation study on rats. Pronounced effects, such as a decrease in fertility and decreased activities of a number of enzymes in the testes were found in rats treated orally with 1000 mg/kg body weight (the only dose level tested), 5 days/week, for 4 months. Besides the effects mentioned, a decreased gestation index was also found, which may indicate early embryonic death due to the induction of a dominant lethal gene in the sperm cells. An increase in mortality and decreased viablility and fertility were found in the F_1 - F_3 generations (not exposed to the carbamate) (Ivanova-Chemishanska & Antov, 1980; Ivanova-Chemishanska et al., 1980). In a second 3-generation rat reproduction study with benomyl, no differences in reproduction or lactation were observed among control and test groups of animals at the highest dietary level of 2500 mg/kg diet. No pathological changes were found in tissues from weanling pups of the F_{3b} litter (Sherman et al., 1975).

(c) Carbendazim

From 3 multigeneration reproduction studies on rats, in which carbendazim was tested at dose levels of up to 10 000 mg/kg diet, no apparent adverse effects on reproduction were found at levels of up to 2000 mg/kg diet. A reduction in average litter weights was observed at dietary levels of 5000 and 10 000 mg/kg (FAO/WHO, 1985a).

(d) Carbaryl

In 3-generation rat studies, carbaryl added to the diet to provide daily doses of up to 10 mg/kg body weight, did not result in any statistically-significant dose-related effects on fertility, gestation, viability of pups, or lactation (Weil et al., 1972).

In other studies by Weil et al. (1972, 1973), groups of rats received carbaryl in the diet at daily doses of 0, 3, 7, 25, or 100 mg/kg body weight by intubation, for 5 days/week, or 0 or 100 mg/kg body weight per day in the diet. Significant effects were found on reproduction (mortality and reduced fertility) in the group receiving 100 mg/kg. A group receiving 200 mg/kg diet did not show any effects.

The influence of carbaryl at dose levels of 2 or 5 mg/kg body weight on the reproductive function of rats was studied by Shtenberg & Otovan (1971). Both dose levels produced adverse effects on the testes and ovaries (atrophic, dystrophic and necrotic changes) and changed the gonadotropic function of the hypophysis. The changes observed increased progressively over the generations. The fecundity of test females fell, the number of still-born offspring increased, and there was a high death rate among young animals.

Carbaryl was fed at dietary levels of 0, 2000, 5000, or 10 000 mg/kg feed to Osborne-Mendel (10 males and 10 females) rats for 3 generations. At the highest dose level, fertility was impaired and no litters were produced from the second mating of the second generation. Marked effects from birth to day 4 were manifested as seen in the survival index. Dose-related decreases in average litter size, number of live-born progeny, number of survivors to day 4, and number weaned were observed.

In a comparative study on gerbils, dietary levels of 0, (40 animals) 2000, 4000, 6000 (each 30 animals), and 10 000 mg carbaryl/kg (18 animals) were tested for 3 generations. No litters were produced from the second mating of the third generation, at the highest dose level. Gerbils appeared to be more sensitive than rats since significant decreases were found consistently at lower dietary levels, but the effects were not clearly dose-related. Effects in both species at 2000 mg/kg diet were still marginal but significant (Collins et al., 1971).

(e) Thiophanate-methyl

Thiophanate-methyl was tested in a 3-generation reproduction study on rats at dose levels of 0, 40, 160, or 640 mg/kg diet. The highest dose level affected growth, but no effects of the compound on reproduction parameters were apparent (FAO/WHO, 1974b).

(f) Carbofuran

A 3-generation reproduction study was conducted on rats (8 males and 16 females per group), with dietary levels of 0, 1, 10, or 100 mg carbofuran/kg diet. The highest dose level resulted in decreased weight gain in the parents and a marked reduction in survival of the young. Lower levels did not have any effects. Follow-up studies with lower dose levels showed that 20 mg/kg diet induced adverse effects on reproduction variables, noted after the first generation (McCarthy et al., 1971; FAO/WHO, 1977b, 1981b).

8.5.2 Endocrine system

(a) Carbaryl

Short-term studies on male and female rats were carried out to study the functioning of the reproductive system under the influence of carbaryl. The dose levels ranged from 2 to 300 mg carbaryl/kg body weight and were administered over periods of 3 - 12 months; in one study, 4 generations were bred. Different criteria were studied, such as reduction of sperm motility, duration of sperm survival, inhibition of spermatogenesis and oögenesis, estrous cycle, induction of degeneration of the germinal epithelium, and the activities of different enzymes in the testes and ovaries. Shtenberg & Rybakova (1968) found marked functional and morphological changes in the endocrine organs of rats administered carbaryl at daily doses of 7, 14, or 70 mg/kg body weight for up to 12 months. It was considered by the authors that the effects of carbaryl on the testes and ovaries might be explained by the observed increase in hypophyseal gonadotropic function. A direct effect on the testes and ovaries cannot be excluded.

Orlova & Zhalbe (1968) studied the effects of carbaryl on white rats. Dose levels of 2.5 or 5 mg/kg body weight were administered. Disturbances in the enzymatic activity in the testes and ovaries, inhibition of spermatogenesis and oögenesis, and decreased fertility were observed in the P generation, and a decrease in viability, in the F_1 generation.

Shtenberg & Otovan (1971) studied this effect of carbaryl in more detail in a 4-generation study. The dose levels studied were 0, 2, and 5 mg/kg body weight, and were applied for 6 months in each generation. The disturbances in the function of the testes and ovaries described above were found at both 2 and 5 mg/kg (Vashakidze, 1965, 1967; Shtenberg & Rybakova, 1968; Shtenberg & Otovan, 1971; Shtenberg et al., 1973). However, the studies reporting adverse effects on spermatogenesis in male rats were not confirmed in studies on mice with carbaryl by Dikshith et al. (1976) or in reproductive performance studies on rats.

(b) Benomyl

A study on the short-term toxicity of benomyl carried out by Torchinsky et al. (1976) showed that the testis of Wistar rats was the organ most affected. When administered over a 12-month period, a dose of 62.5 mg/kg body weight showed borderline effects; 12.5 mg/kg body weight did not show an effect. A recent study on Sprague Dawley rats receiving 10 daily treatments of 0, 200, or 400 mg benomyl/kg body weight, by gavage, showed a significant decrease in the total epididymal sperm counts and sperm concentrations in the vas deferens at both levels. At the higher dose level, slight to severe hypospermatocytogenesis and generalized hypospermatogenesis were observed (Carter & Laskey, 1982; FAO/WHO, 1985a). Pastushenko (1983) studied the gonadotoxic effects of benomyl in rats exposed to dose levels of 0.5, 1.2, 5.0, or 12.4 mg/m^3 air. The 2 highest dose levels showed dose-related changes in the gonads, i.e., disturbances of the morphological and functional states. At 1.2 mg/m^3, only morphological changes were observed. No effects were found with 0.5 mg/m^3.

(c) Carbendazim

When Sprague Dawley rats were fed diets containing 0, 2000, 4000, or 8000 mg carbendazim/kg diet for 28 days, degeneration of testicular tissue was observed at 4000 and 8000 mg/kg diet. Spermatogenesis and oögenesis were affected at these feeding levels. Changes were also found in the organ weights relative to body or heart weight, changes in the liver being observed at 2000 mg/kg diet (FAO/WHO, 1974b).

No effects on testicular weights and morphology were found in a study on groups of ChR-CD rats fed carbendazim in the diet for 90 days at dose levels of 0, 100, 500, or 2500 mg/kg (FAO/WHO, 1985a).

8.5.3 Embryotoxicity and teratogencity

A number of carbamate esters have been tested for teratogenicity with the chicken embryo bioassay (Marliac, 1964; Ghadiri & Greenwood, 1966; Khera, 1966; Ghadiri et al., 1967; Rybakova, 1968; Olefir & Vinogradova, 1968; Proctor et al., 1976; Moscioni et al., 1977). However, this test is not considered of relevance in testing for teratogenicity, and the results will not be discussed in this document.

(a) Propoxur

Roszkowski (1982) studied the effects of propoxur on intrauterine development and on the state of ossification of rat fetuses. An embryotoxic effect was found at a dose of 20% of the LD_{50}, administered on days 7 - 19 of the gestation period; 11.9% of anomalies occurred in the fetuses of treated animals compared with 6.3% in the controls. Decreased weight was also found in the fetuses. Ossification was disturbed, but this was not related to the period of gestation during which the substance was administered. A decrease in the concentrations of calcium and magnesium was found in the serum of mothers whose fetuses had shown bone anomalies.

(b) Benomyl

In a teratogenicity study, pregnant ChR-CD rats were fed a diet containing 0, 100, 500, 2500, or 5000 mg benomyl/kg from day 6 to day 15 of gestation; no effects were found on the outcome of pregnancy or embryonal development (Sherman et al., 1975). Hazleton Laboratories (1968) cited in FAO/WHO (1985a) reported similar negative results for rabbits, at dose levels up to 500 mg benomyl/kg diet administered on days 8 - 16 of gestation.

Torchinsky (1973) studied the influence of carbaryl (106 mg/kg body weight, daily, from day 1 to day 20 of pregnancy) and benomyl (single dose of 145 or 250 mg/kg body weight on day 12 of pregnancy) in relation to the protein content of the diet. The animals were kept on diets containing 5, 10, 19, or 35% of protein either for one month prior to mating and during the entire pregnancy or only during pregnancy. The teratogenic and embryotoxic actions of benomyl and carbaryl did not differ in groups where the females received diets containing 5, 10, or 19% protein from the beginning of pregnancy. An increased post-implantational lethality for the embryos was noted in females kept on a diet with 5% protein from one month before conception. When these animals were exposed to carbaryl, the weight of 20-day-old fetuses declined considerably, yet the severity of the teratogenic action of benomyl did not increase under these conditions. The teratogenic action of benomyl and the embryotoxic action of carbaryl were less severe in animals fed diets containing 10% of protein from one month before conception than in the groups kept on diets containing 19 or 35% protein. The results of a few other studies have indicated that benomyl causes embryotoxic and teratogenic effects in both rats and mice when administered orally (by gavage) at dose levels at 62.5 mg/kg body weight (rat) and 100 mg/kg body weight (mouse) or more. At lower dose levels (31.2 mg for the rat and 50 mg/kg for the mouse), no teratogenic effects were observed. A broad spectrum of malformations such as exencephaly, hydrocephaly, and meningocephaly, and disturbance of the CNS were found (Ruzicska et al., 1975; Petrova-Vergieva, 1979; Kavlock et al., 1982).

(c) Carbaryl

A number of teratogenicity studies carried out with carbaryl gave contradictory results. Weil et al. (1972) administered up to 500 mg carbaryl/kg body weight to rats. No teratogenic effects or effects on fertility or gestation were seen at this high dose level; however, the mean body weight gain of the females during pregnancy was decreased. Many pups died before weaning. No effects were found with administration of carbaryl at 100 mg/kg body weight (Weil et al., 1972).

Oral administration of carbaryl to pregnant rabbits and hamsters at doses of up to 250 mg/kg body weight, by intubation, did not cause any teratogenic effects or dose-related fetal mortality (Weil et al., 1973).

Dougherty et al. (1975), cited in FAO/WHO (1974b), did not observe any teratogenic effects in monkeys given doses of carbaryl of 0, 0.2, 2, or 20 mg/kg body weight from day 18 to day 40 of gestation. This study indicated that even the highest dose level did not induce a reproductive hazard.

Vashakidze (1965, 1967) found reduced reproduction and abnormal fetal changes with intubated dose levels of 100 mg/kg body weight and higher. A level of 50 mg/kg produced negative results. Carbaryl, administered to guinea-pigs orally during organogenesis (10 consecutive days between days 11 and 20 of gestation) at 300 mg/kg body weight, induced increased maternal and fetal mortality and offspring with axial skeletal defects, such as cervical vertebrae fusion. No abnormal fetal changes were produced in hamsters at the sub-lethal levels of 125 and 250 mg/kg body weight. In rabbits, levels ranging from 50 to 200 mg/kg body weight did not produce any effects on mortality or morbidity, and no abnormal fetal changes were seen (Robens, 1969).

Smalley et al. (1968) conducted studies on pregnant beagle dogs fed a diet containing 0, 3.125, 6.25, 12.5, 25, or 50 mg carbaryl/kg body weight from mating to the termination of weaning. Effects observed included a number of animals with dystocia (difficult births) due to atonic uterine musculature, an apparent contraceptive effect at the highest dose level, and teratogenic effects (different types of severe defects) at all but the lowest dose level.

The effects of carbaryl on rat reproduction and guinea-pig teratology were compared by Weil et al. (1973). A 3-generation reproduction study on rats included groups receiving carbaryl at maximum daily doses of 200 mg/kg diet. Other groups concurrently received daily oral intubation doses as high as 100 mg/kg body weight. In the guinea-pig teratology study, maximum levels of 300 mg/kg diet or 200 mg/kg body weight (intubation) were given. The concurrent administration of carbaryl by daily oral administration or by dietary inclusion did not result in either teratological or mutagenic effects in the rat or teratological effects in the guinea-pig, at the maximum dosage levels. However, the groups receiving maximum levels orally showed severe toxic effects in both studies compared with little or no effect with oral administration of 25 mg/kg in rats and 100 mg/kg in guinea-pigs.

The authors concluded that the severe effects found in rats by Shtenberg & Otovan (1971) following repeated oral intubation of carbaryl at 2 mg/kg body weight could not be confirmed, even at the maximum oral intubation level of 100 mg/kg body weight.

(d) Methomyl

When rabbits were fed dietary levels of 0, 50, or 100 mg methomyl/kg on days 8 - 16 of gestation, no skeletal or visceral abnormalities were seen in the fetuses (Kaplan & Sherman, 1977).

(e) Carbofuran

Teratogenicity studies have been carried out on albino rats and albino rabbits with carbofuran. In rats, dose levels of 0, 0.1, 0.3, or 1.0 mg/kg body weight were administered throughout gestation (days 6 - 15 of pregnancy) and, in rabbits, levels of 0, 0.2, 0.6, or 2.0 mg carbofuran/kg body weight were administered from day 6 to day 18 of gestation. In both animal species, the highest dose level showed clear signs of poisoning. However, pregnancy, viability, fetal body weights, and all the other examinations did not show any effects. All the progeny were normal. No teratogenic effects were seen in rats or rabbits (McCarthy et al., 1971; FAO/WHO, 1981b).

(f) Carbendazim

Carbendazim has been shown to be embryotoxic at dose levels of approximately 19 mg/kg body weight or more. This effect was not seen at 10 mg/kg body weight (Delatour & Richard, 1976).

8.6 Mutagenicity and Related End-Points

There is little evidence, if any, to suggest that methyl carbamate compounds are mutagenic. An evaluation of the mutagenic potential of carbaryl, methomyl, propoxur, carbofuran, and pirimicarb, using a variety of microorganisms as indicator strains, including strains of *Salmonella typhimurium*, *Escherichia coli*, *Saccharomyces cerevisiae*, and *Bacillus subtilus*, produced negative results. These studies showed that carbamate compounds have little potential for presumptive gene mutations in higher animals.

Anderson et al. (1972) examined barban, chloropropham, propham, and swep for their ability to induce point mutations in 8 histidine-requiring strains of *S. typhimurium*. None showed induced mutations. This study was carried out in the absence of any metabolic system.

The chemical structure of benomyl probably enables it to act both as a base analogue for DNA (benzimidazole ring incorporates into nucleic acids instead of guanine) and as a spindle poison (carbamate groups) (Seiler, 1975, 1976b; FAO/WHO, 1985a). The binding of benomyl to microtubular proteins disturbs the mitotic spindle, thus causing malsegregation of chromosomes during the mitotic anaphase (Morris, 1980, cited by Finnish National Board of Health Toxicology Group, 1982). The results of several studies suggest that benomyl is a strong antimitotic agent which, even in low concentrations, induces non-disjunction of single chromosomes leading to aneuploidy (FAO/WHO, 1985a).

The induction of low-frequency forward mutations by carbendazim in *S. typhimurium* strain LT-2 was reported by Seiler (1972). Reversions reported to occur in *S. typhimurium* strains His G46 and TA 1530 were attributed to base-substitution mutations caused by carbendazim.

Another study was made with the WP2 uvrA repair-deficient strain of E. coli and the TA 1535 repair-deficient strain of S. typhimurium. Mutagenic activity was detected for benomyl in the former, but not in the latter or in TA 1538 (Kappas et al., 1976). No effects were found in the CM-611 strain of E. coli, deficient in either the excision repair pathway or the lex A+-dependent misrepair pathway. In studies carried out recently, benomyl was negative in all S. typhimurium strains used, even though, in some of the repeats, significant increases in the number of revertants were observed in TA 1535 strain. Carbendazim was clearly mutagenic in the TA 98 strain with metabolic activation (S-9 mix) and slightly mutagenic in the TA 1535, TA 1537, and TA 100 strains. A positive result for carbendazim in the TA 1537 strain was observed, also without metabolic activation (Haskell Laboratories, cited by the Finnish National Board of Health, 1982; FAO/WHO, 1985a).

There is evidence that carbendazim and related compounds are antimitotic agents and cause mitotic arrest, mitotic delay, and a low incidence of chromosome damage in SPF Wistar rats and mammalian cell lines. The mutagenic effects of carbendazim in bacteria are probably due to base substitutions, resulting from the incorporation of small quantities of benzimidazole into nucleic acids. The effects on mammalian cells in vivo and in vitro indicate that carbendazim causes chromosome damage, which may result from this type of mutagenic activity, but the effects on mitosis suggest that, in higher organisms, the main genetic effect is to cause non-disjunction (Styles & Garner, 1974).

An abnormal number of micronucleated erythrocytes have been observed in the bone marrow of mice treated orally with carbendazim at 100 mg or more/kg body weight. No effects were seen at 50 mg/kg. Benomyl at 1000 mg/kg body weight produced a slight increase in micronucleated erythrocytes.

Carbendazim did not induce chromosome breakage in Chinese hamster bone-marrow cells at oral doses as high as 1000 mg/kg body weight (Seiler, 1976a). It was concluded that carbendazim is not a chromosome-breaking agent, and its micronucleic action is attributable to antimitotic activity caused by inhibition of spindle formation or spindle function.

An evaluation of the mutagenic properties of benomyl has been completed and reported by the US Environmental Protection Agency Pesticide Genotoxicology Program. Benomyl increased the mutation frequency in a concentration-related manner in L 51784 mouse lymphoma cells and induced sister chromatid exchange in Chinese hamster ovary (CHO) cells. Benomyl was also positive at all dose levels tested in the in vivo micronucleus test with mice. The results of the spot test conducted on mice support the idea that carbendazim (oral application of 100 - 300 mg/kg body weight to the mother on the 10th day following conception) may be mutagenic in mammalian cells due to a direct action on DNA (Fahrig & Seiler, 1979).

A dominant lethal study to evaluate the potential mutagenic effects in the reproductive cycle of the male mouse was conducted with propoxur at dose levels of 0, 2.5, or 5 mg/kg body weight. For a period of 6 weeks, 3 virgin females were exposed to each male at weekly intervals and reproduction indices recorded. The administration of propoxur proved to have a transient effect on the reproductive capability of the males but there was no sign of early resorption indicating a mutagenic effect (FAO/WHO, 1974b).

DNA-repair tests (with both isolated rat and mouse hepatocytes) were conducted using benomyl and carbendazim. All tests showed negative results for the induction of DNA repair (Haskell Laboratories, 1981a,b cited in FAO/WHO, 1985a).

The Finnish National Board of Health (1982) summarized the mutagenicity data on benomyl, carbendazim, and thiophanate-methyl (Table 7) and concluded that the results were sometimes contradictory. However, positive mutagenicity results from point mutation tests and chromosome aberration tests are well documented, especially in higher organisms, and it can be concluded that carbendazim fungicides should be considered as weak mutagenic compounds.

Without going into detail, it is known that a large number of *N*-nitroso derivatives of methyl carbamates are potent mutagens. The *N*-nitroso derivatives of aldicarb, propoxur, carbofuran, trimethacarb, and methomyl caused numerous single-strand breaks in normal human skin cells. The data suggested that human cellular DNA *in vivo* was irreversibly altered resulting in numerous alkali-sensitive bonds (Blevins et al., 1977).

A similar result was obtained with *N*-nitroso carbaryl. This compound showed strong mutagenic activity in 2 bacterial systems, *E. coli* and *Haemophilus influenzae*. Using *S. typhimurium*, *N*-nitroso carbaryl acted as a potent base-pair substition mutagen and showed a relatively mild frameshift activity. The parent compound, carbaryl, gave negative results (Regan et al., 1976; FAO/WHO, 1982b).

Uchiyama et al. (1975) tested the nitroso derivatives of BPMC, propoxur, hoppcide, isoprocarb, MPMC, and MTMC, in the back mutation test with *E. coli* B/r WP2 try$^-$ and in the rec-assay method with *B. subtilus* strains. The *N*-nitroso derivatives of all these *N*-methylcarbamates showed potent mutagenic activity.

Table 7. A summary of the mutagenicity data on benomyl, carbendazim (MBC), and thiophanate-methyl

Point mutations	Genetic change	
	Chromosomal aberrations	Non-disjunction
Positive results +	+	+
Salmonella typhimurium (Seiler, 1972)	rat fetus (Ruzicska et al., 1975)	
Salmonella typhimurium (Haskell Laboratories, 1982 cited in FAO/WHO, 1985a)	rat bone marrow (Styles & Garner, 1974)	rat bone marrow (Styles & Garner, 1974)
		mouse bone marrow (Seiler, 1976a)
Escherichia coli (Kappas et al., 1976)		Aspergillus nidulans (Kappas et al., 1974; Bignami et al., 1977)
Fusarium oxysporum (Dassenoy & Meyer, 1973)		
Mouse/spot-test (Fahrig & Seiler, 1979)	Chinese hamster ovary (SCE) (SRI International, 1980, cited in FAO/WHO, 1985a)	
	micronucleus in vivo/mouse (SRI International, 1980, cited in FAO/WHO, 1985a)	

Table 7 (contd).

Negative results -	-	-
Streptomyces coelicolor (Carere et al., 1978)	rat bone marrow (Ruzicska et al., 1975)	dominant lethal/rat (Sherman et al., 1975)
Salmonella typhimurium (Fiscor et al., 1978) (Carere et al., 1978)	Chinese hamster bone marrow (Seiler, 1976a)	dominant lethal/mice (Makita et al., 1973)
Salmonella typhimurium (Haskell Laboratories, 1981; cited in FAO/WHO, 1985a)	lymphocyte culture-cytotoxicity obs. (Lamb et al., 1980)	
Aspergillus nidulans (Bignami et al., 1977)	benomyl-workers (Ruzicska et al., 1975)	
Drosophila/recessive lethal - sterility obs. (Lamb & Lilly, 1980)	hepatocyte culture/DNA-repair rat, mouse (Haskell Laboratories, 1981, cited in FAO/WHO, 1985a)	
	human lymphocyte culture (Gupta & Legator, 1975, cited in FAO/WHO, 1985a)	

From: The Finnish National Board of Health (1982) FAO/WHO (1985a).

8.7 Carcinogenicity

8.7.1 General

The mechanism of the carcinogenic action of ethylcarbamate (urethane) has not yet been established, but its structure seems to be optimal. All changes seem to reduce activity, particularly when the ethyl group is replaced with larger side chains. The addition of alkyl groups to the nitrogen also reduces activity (Mirvish, 1968). It is not so clear why urethane is carcinogenic, whether it is the proximal toxin or whether it has to be converted to an active metabolite. The low carcinogenic potential of esters of N-substituted carbamic acids is possible.

In 1976, a Working Group of the International Agency for Research on Cancer (IARC, 1976) reviewed the available data on a number of carbamate pesticides included carbaryl, propham, chloropropham, and mexacarbate.

Carbaryl was tested in a number of carcinogenicity tests some of which were not relevant or were of inadequate design. The rat study (40 per group), in which carbaryl was administered at dose levels of 0, 50, 100, 200, or 400 mg/kg diet for 2 years, did not give any indication of an increase in tumour incidence (Carpenter et al., 1961; Innes et al., 1969; Andrianova & Alekseev, 1970; IARC, 1976).

Groups of 48 male and 48 female CD-1 mice were given 0, 100, or 400 mg carbaryl/kg diet. After 80 weeks, the tumour incidence between the groups was similar (Mellon Institute, 1963 cited in FAO/WHO, 1965b).

Mexacarbate was administered orally to 2 strains of mice. The design of this study was inadequate (only a small number of animals), but an increased incidence of lung adenomas was shown in both strains, together with a slight increase in liver tumours in a number of males of one strain. According to IARC (1976), it was not possible to make an evaluation of the carcinogenicity of mexacarbate on the basis of this study by Innes et al. (1969).

Propham and chlorpropham were administered via the oral route in a number of carcinogenicity tests on mice, rats, and hamsters. One or both compounds were also tested on mice and/or rats via the subcutaneous, intraperitoneal, or intrapleural route. The design of some of these studies was inadequate. Propham and chloropropham were tested for initiation properties, because the ethyl carbamates are related to the well-known carcinogen ethylcarbamate (urethane), which is an initiator of skin tumours. There was no evidence that propham and/or chloropropham were carcinogenic after oral, subcutaneous, intraperitoneal, and intrapleural application. However, these two compounds acted as weak initiators in the two-stage carcinogenicity studies on mice. A second study by the same authors did not confirm the results (Van Esch et al., 1958; FAO/WHO, 1965b; Innes et al., 1969; Van Esch & Kroes, 1972; IARC, 1976).

Long-term studies on rats did not provide any evidence of carcinogenic activity when propoxur was fed in the diet for 2 years at levels of 0, 250, 750, 2000, or 6000 mg/kg diet. Increased liver weight was found at the highest dose level (6000 mg/kg) in males and in the three highest dose levels in females. This liver change was not reflected in liver function tests, clinical chemical examinations, or histological changes (FAO/WHO, 1974b). Similarly, long-term and 2-year studies on rats, dogs, mice given carbofuran or pirimicarb did not indicate any carcinogenic potential. The dose levels tested were 175 mg/kg diet for rats and 1.8 mg/kg body weight for dogs for pirimicarb, 100 mg/kg diet for rats, and 500 mg/kg diet for mice (24 months in mice) for carbofuran (FAO/WHO, 1977b, 1981b). On the basis of a variety of test protocols, these carbamate compounds did not show any significant carcinogenic potential.

Benomyl was tested for carcinogenicity in CD-1 mice, at dose levels of 0, 500, 1500, or 5000 mg/kg diet (approximately equivalent to 0, 50, 150, or 500 mg/kg body weight). A significant dose-dependent increase in the incidence of primary hepatocellular adenomas and carcinomas in female mice was seen. Furthermore, histopathological changes occurred in a number of organs (Haskell Laboratories, 1981 cited in FAO/WHO, 1985a).

Low dose levels of thiophanate-methyl (0 - 640 mg/kg diet, equivalent to approximately 0 - 64 mg/kg body weight) were tested for carcinogenicity in ICR-SCL mice. This study showed that ingestion of 640 mg thiophanate-methyl in the diet by a mouse strain susceptible to various tumours did not affect carcinogenic activity. In another study on rats, dose levels of 0, 10, 40, 160, or 640 mg/kg diet did not have any apparent effects on the occurrence of tumours; the only significant effects appeared to be decreased growth at the highest dose levels (Hashimoto & Fukuda, 1972, cited in FAO/WHO, 1974b).

The influence on the testes of ChR-CD rats of carbendazim administered at dose levels of 0 - 10 000 mg/kg diet was studied over a period of 2 years. However, the incidence of testicular changes was so high in the two control groups that no opinion could be given on the possible influence of carbendazim on the testes (Haskell Laboratories, 1977 cited in FAO/WHO, 1985a).

Carbendazim at dose levels of 0, 150, 300, or 1000 mg/kg diet (this dose level was increased to 2000 mg/kg at week 4 and to 5000 mg/kg at week 8 for the remainder of the study) was administered to SPF-Swiss mice. The duration of the study was 80 weeks. At the highest dose level, neoplastic nodules were found in the livers of female mice. In males, there was an increased incidence of hepatoblastomas (CIVO/TNO, 1976 cited in FAO/WHO, 1985a).

A 2-year feeding study on CD-1 mice was conducted using carbendazim at dose levels of 0, 500, 1500, or 7500 mg/kg diet (only females completed the 2-year period, the 7500 mg/kg level for males was decreased to 3750 mg/kg after the first 15 months of the study and the final sacrifice of this group was antedated by 7 months). A significant increase in hepatocellular carcinomas occurred at 1500 mg and 7500 mg/kg in female mice and at 1500 mg/kg in male mice (at the highest dose level for males, there were too few survivors to judge the effects) (Haskell Laboratories, 1982; FAO/WHO, 1985a).

Other studies demonstrated that carbamates in the presence of nitrite could be converted to N-nitroso derivatives, which may be (partly) responsible for the "carcinogenic activity of the carbamates".

Studies in which carbendazim was administered to Swiss mice, twice, at 250 mg/kg body weight, intragastrically, weekly, in combination with 5000 mg sodium nitrite/litre drinking-water caused a slight increase in the incidence of lymphosarcomas (Börzsönyi & Csik, 1975; Börzsönyi & Pinter, 1977).

The results of an in vitro study of Beraud et al. (1979) (IARC, 1976) suggested that nitrosocarbaryl could be produced in rat gastric juice from carbaryl.

The proposal that carbamate pesticides may be converted to N-nitroso compounds under the acid conditions of the stomach is based on the results of in vivo nitrosation studies with secondary amines and other compounds in the presence of nitrite (originating, for instance, from certain vegetables and present in saliva) (Tannenbaum et al., 1974). Elespuru & Lijinsky (1973), Eisenbrand et al. (1974, 1975a,b), Elespuru et al. (1974), Ungerer et al. (1974), Quarles & Tennant (1975), Lijinsky & Taylor (1976), and Oliver (1981) have shown that certain pesticides, such as carbaryl and dithiocarbamates, can be nitrosated in vivo.

As pointed out by IARC (1976), the extrapolation of findings in experimental animals to man is complicated by many factors. It is relatively easy to show that N-nitroso derivatives can be formed and that these are mutagenic and/or carcinogenic. However, the crucial information is the quantity produced in man under the prevailing conditions. The concentration of both reactants, differences in pH, the influence of competing reactions, and the presence of accelerators and inhibitors are all important. In addition to these difficulties in defining potential human exposure, the susceptibility of man has to be compared with that of experimental animals. General considerations on *N*-nitrosatable pesticides are described in IARC (1983).

8.8. Special Studies

Adequate characterization of the neurobehavioral effects of carbamates is still incomplete or not available. In general, when given in single doses the ChE-inhibiting carbamates decrease activity and schedule-controlled behaviour, interfere with avoidance and escape performance, and decrease the amplitude of a variety of electrophysiological measures. The possible interference with learning and memory by the carbamates has been of interest because of the hypothesized role of the cholinergic system in these functions. However, the evidence that these compounds interfere with learning and memory is equivocal (Miller, 1982).

Carbamates interfere with shock-motivated behaviour including discrete-trial and one-way avoidances as well as shock-induced suppression. Furthermore, these compounds decrease responsiveness to shock (Sideroff & Santolucito, 1972).

The effects of the ChE-inhibiting carbamates have been evaluated in man. However, these studies have been mainly concerned with correlating ChE inhibition and overt symptoms. It is assumed that, if the clinical indicators (such as ChE levels) are normal after pesticide exposure, or if symptoms associated with cholinergic overstimulation are absent, then no risk is involved (Miller, 1982).

9. EFFECTS ON MAN

The health hazards from overexposure to carbamate insecticides occur primarily as a result of the inhibition of the enzyme ChE. ChE inhibition leads to the accumulation of endogenous ACh and other choline esters in the organism. This produces signs and symptoms as specified in section 9.3.

9.1 General Population Exposure

9.1.1 Acute toxicity: poisoning incidents

Carbamate pesticides have occasionally been used in cases of suicide. Fatal poisonings have occurred with carbaryl, propoxur, and mexacarbate. Farago (1969) described a case involving a 39-year-old man who drank 1/2 litre of carbaryl. He entered the hospital 1 1/2 h later and received gastric lavage, but developed serious respiratory depression. His condition improved somewhat after 4 injections of atropine at 1/2-h intervals. Three hours after the carbaryl ingestion, 250 mg of 2-PAM intravenous was administered. His lung condition rapidly worsened and he died. Administration of 2-PAM is contraindicated for cases of carbaryl poisoning, as it may increase toxicity rather than reactivate the inhibited enzyme and, in this case, it was concluded that the fatal outcome, though unavoidable, was hastened by 2-PAM administration.

Reich & Welke (1966) reported a fatal case of mexacarbate poisoning involving a 17-year-old nursery employee who ingested approximately 55 g mexacarbate in the form of a 22% formulation. The young man, found unconscious, had pinpoint pupils and an irregular heartbeat. Emergency procedures at the hospital briefly restored regular heart rhythm, but brachycardia developed later with recurrent heart failure. The patient died about 4 - 4 1/2 h after the ingestion of the mexacarbate.

Two severe cases of propoxur poisoning following ingestion of 150 - 200 ml of commercial Baygon formulation were described by Bomirska & Winiarska (1972). One individual responded to treatment and recovered but the other died after failing to respond to treatment with atropine and 2-PAM and toxogonin given intravenously. A normal ChE suggested spontaneous reactivation of this enzyme.

9.1.2 Effects of short- and long-term exposure

Propoxur was tested for its effectiveness in controlling the mosquito vectors of malaria in Iran. The insecticide was sprayed inside houses in a 300 km^2 area in which 11 000 people lived in 33 villages. Sixty-nine out of the 11 000 inhabitants complained of mild cholinergic symptoms, when they entered their houses during the spraying, contrary to instructions. The symptoms were headache, nausea, sweating, and vomiting. The symptoms subsided in 2 - 3 h. Similar experiences were reported in other villages sprayed with propoxur or with carbaryl (Vandekar et al., 1968).

9.1.3 Controlled human studies

Human volunteers tolerated single oral doses of aqueous solutions of aldicarb at doses of 0.025, 0.05, or 0.1 mg/kg body weight, with only mild depression of whole blood-ChE in all cases. Only the persons receiving 0.1 mg/kg presented symptoms of acute cholinergic stress (i.e., nausea, sweating, pinpoint pupils, salivation) (FAO/WHO, 1982b).

Groups of male volunteers (5 or 6 per group) were administered daily doses of 0.06 and 0.13 mg carbaryl/kg body weight, orally, for 6 weeks. No subjective or objective changes were observed, except a slight reversible decrease in the ability of the kidneys to reabsorb aminoacids in the group at the highest dose level (Wills et al., 1968).

Thirty-eight urban volunteers were monitored for carbaryl exposure during the summer. All volunteers were involved in the application of carbaryl incidental to their employment or leisure activities. Exposure of clothing, skin, with and without gloves, were measured. The maximum dermal exposure recorded was 2.86 mg/kg per h, the maximum air concentration was 0.28 mg/m^3. Only a small mean decrease in serum- and erythrocyte-AChE activity was found (Gold et al., 1982). Comparable results were obtained by Leavitt et al. (1982) in a study on 2 groups of professional pesticide spray operators with higher dermal exposure, spraying trees with carbaryl.

A 42-year-old male volunteer took a single oral dose of 135 mg propoxur (1.5 mg/kg body weight) about 2 h after breakfast. Plasma- and erythrocyte-ChE activities were determined every 15 min for the first 2 h after ingestion and then every hour thereafter. Pulse and blood pressure were recorded. A rapid drop in erythrocyte-ChE activity was observed, reaching the lowest level of 27% of normal within 20 min. ChE activity gradually recovered, reaching normal levels about 2 h after ingestion. Plasma-ChE activity was not affected. In another study (single dose of 0.36 mg/kg body weight), the lowest erythrocyte-ChE activity occurred within 10 min together with short-lasting stomach discomfort, blurred vision, nausea, and facial perspiration. Twenty min after ingestion, the volunteer's pulse rate and blood pressure were increased. Pronounced nausea accompanied by repeated vomiting and profuse sweating occurred until approximately 45 min after the ingestion of propoxur. From then on, poisoning symptoms diminished, blood pressure and pulse returned to pre-ingestion values. The disappearance within about 2 h of the poisoning symptoms was in line with the return of normal erythrocyte-ChE activity. Forty-five percent of the ingested dose of propoxur was excreted as the metabolite, 2-(1-methylethyl)phenol, within 24 h of ingestion (Vandekar et al., 1971; FAO/WHO, 1974b).

In another study, 5 doses of propoxur, at 0.15 or 0.20 mg/kg body weight, were administered to volunteers at 1/2-h intervals. In this study, a symptomless depression of erythrocyte-ChE to 60% of normal was observed over the course of the dosing. The enzyme activity recovered to normal levels soon after termination of the dosing. From the above studies, it was concluded by the authors that an oral dose of 0.36 mg propoxur/kg body weight may be sufficient to induce initial cholinergic symptoms, while larger doses may be tolerated without symptoms if the administration is gradually (divided doses) (Vandekar et al., 1971).

9.2 Occupational Exposure

9.2.1 Acute toxicity: poisoning incidents

Tobin (1970) reported 3 cases of persons poisoned with carbofuran. Two were formulation plant employees, preparing 10% granules and the third was an entomologist who began to feel uncomfortable while weighing a 50% water-dispersible powder formulation. The 2 formulators developed similar symptoms indicative of carbamate poisoning: profuse sweating, weakness, blurred vision, and nausea. Both were taken to their respective physicians about 3 h after the onset of symptoms. One physician administered atropine; the other, noting the regression of the symptoms, did not administer atropine. The patient who received the atropine (0.02 g im) was fully recovered in 30 min. The untreated patient recovered over the course of 2 - 3 h. The third case, the entomologist experienced mild discomfort that regressed without atropine administration in 4 - 6 h.

Richardson & Batteese (1973) described the case of a spray-plane co-pilot overexposed to mexacarbate during an spraying operation. The cause of the overexposure was a pin-hole leak in a high pressure pump line, which emitted a fine aerosol of the mexacarbate formulation into the fuselage. The co-pilot developed the classical signs of ChE poisoning. Toxic symptoms progressed to paralysis of the extremities before he could be treated. In hospital, atropine sulfate was immediately administered intravenously followed by a second injection subcutaneously 40 min later. The symptoms subsided quickly. ChE activity in blood drawn after the first atropine injection was severely depressed, but returned to normal after 3 days.

Minor and transient cholinergic symptoms were experienced by spraymen applying propoxur formulations inside houses for testing this insecticide against mosquito vectors of malaria in villages in El Salvador (Quinby & Babione, 1966) and Iran (Vandekar et al., 1968; Montazemi, 1969). A pronounced fall in blood-ChE activity during the work and a distinct recovery after exposure ceased was established as a daily pattern of activity of this enzyme, erythrocyte-ChE being more sensitive to propoxur than plasma-ChE. No cumulative inhibition of ChE during the 6-week exposure was seen. The investigators noted that the exposed spraymen and a few inhabitants recovered rapidly when removed from the source of exposure and when treated with atropine.

Vandekar et al. (1968) commented that most spraymen who became poisoned had heavy skin contamination, and had not cleaned their skin sufficiently after spraying. The inhabitants entered the house while it was being sprayed. Moreover, it was noted that respirators did not offer protection against intoxication under these circumstances and that dermal absorption accounted for most of the absorbed dose.

No adverse health effects due to carbaryl exposure (except for one case of skin rash) were observed in a study conducted by the WHO Insecticide Testing Unit in Southern Nigeria. Ten men (out of 105 sprayers) spraying over a single 6-h period and 95 villagers who resided in the sprayed houses, were examined. The sprayers wore overalls, hats, and rubber boots during the entire 6 h of spraying, but wore masks for the first hour only. Plasma-ChE activity as well as urinary metabolites of carbaryl were determined before and after spraying. A slight (average 5%) reduction in plasma-ChE activity was reported for all 10 sprayers, the day after the carbaryl application. Inhabitants of sprayed houses showed no inhibition of plasma-ChE activity, 1 week after the spraying. A slight increase in 1-naphthol was found in the urine of the sprayers only on the sixth day after the exposure (Vandekar, 1965).

In a study on 19 agricultural workers in the USSR, whole blood-ChE activity was measured before and after 4 to 6-h exposures to airborne carbaryl for 3 - 4 days. Significant ChE inhibition was found in men exposed to a mean airborne carbaryl concentration of up to 4 mg/m^3. No objective signs of ill health were observed. No changes were measured at 0.7 mg/m^3 (Yakim, 1967).

A case of acute intoxication of a woman with a total dose of 60 mg carbofuran was described by Izmirova et al. (1981).

Slight ChE inhibition was found, but within 72 h the patient recovered completely.

The above reports identify systemic ChE poisoning as the major hazard associated with occupational exposure to carbamate pesticides. However, other publications indicate that certain of these compounds can also irritate the exterior membranes of the body. Tobin (1970) described 2 cases of field workers that illustrate direct effects on the eye, (irritation and blurred vision) during carbofuran spraying operations. These 2 persons did not developed symptoms of systemic poisoning.

Vandekar (1965) reported skin rash in a sprayman accidentally splashed with a carbaryl formulation. Although carbamate compounds have not generally been implicated as the cause of dermatitis or allergic skin reactions, Vandekar (1965) suggested that certain individuals could develop this after unusually heavy exposure.

9.2.2 Effects of short- and long-term exposure

Little is known about the short- and long-term effects of occupational exposure to carbamate pesticides. The few reports that discuss this aspect do not implicate long-term exposure to the carbamate compounds as a cause of human disease. Vandekar et al. (1968) did not observe any evidence of cumulative toxicity among spraymen applying propoxur formulations during a 6-week operation. No long-term biochemical changes were reported among New Jersey farmers and spraymen compared with a minimally exposed control group. It was noted that significantly higher incidences of hearing and eye problems, chronic cough, dizziness, headaches, and hypertension occurred in the occupationally exposed group. However, the various pesticides to which this group was exposed were not tabulated (Kwalick, 1971).

Chromosome studies on the blood cultures of 20 workers engaged in the production of benomyl did not reveal any increased frequency of structural chromosome aberrations (Ruzicska et al., 1975).

In a survey of occupationally acquired disease in workers at a pesticide plant, 11 out of 102 workers had been hospitalized for illness related to chemical exposure. The commonest cause of hospitalization was intoxication with the ChE inhibitor methomyl. On clinical evaluation, 5 out of 11 packaging workers, with the highest exposure to methomyl had experienced blurred vision or pupillary constriction (Morse & Baker, 1979).

9.2.3 Epidemiological studies

Epidemiological studies on persons occupationally exposed exclusively (or at least primarily) to carbamate pesticides are not available. Such studies are needed before the long-term effects of this class of pesticides on human beings can be adequately evaluated.

9.3 Signs and Symptoms of Acute Intoxication by Carbamates

The clinical picture of carbamates intoxication results from accumulation of ACh at nerve endings. Extensive description of the syndrome is given in several major references (Namba et al., 1971; Kagan, 1977; Taylor, 1980). The signs and symptoms can be categorized into the following 3 groups:

(a) Muscarinic manifestations

- increased bronchial secretion, excessive sweating, salivation, and lachrymation;

- pinpoint pupils, bronchoconstriction, abdominal cramps (vomiting and diarrhoea); and

- bradycardia.

(b) Nicotinic manifestations

- fasciculation of fine muscles (in severe cases, diaphragm and respiratory muscles also involved); and

tachycardia.

(c) Central nervous system manifestations

- headache, dizziness, anxiety, mental confusion, convulsions, and coma; and

depression of respiratory centre.

All these signs and symptoms can occur in different combinations and can vary in onset and sequence, depending on the chemical, dose, and route of exposure. The duration of symptoms is usually shorter than that observed in organophosphorus poisoning. Mild poisoning might include muscarinic and nicotinic signs only. Severe cases always show central nervous system involvement; the clinical picture is dominated by the respiratory failure sometimes leading to pulmonary oedema due to the combination of the above-mentioned symptoms.

Clinical diagnosis is relatively easy and is based on:

(a) medical history and circumstances of exposure;

(b) presence of several of the above-mentioned symptoms, in particular, bronchoconstriction and pinpoint pupils not reactive to light; pulse rate is not of diagnostic value, because the AChE effects on the heart reflect the complex innervation of this organ; on the other hand, since changes in the conduction and excitability of the heart might be life-threatening, monitoring should be performed.

(c) confirmation of diagnosis is made by measurement of AChE in RBC or plasma-pseudoChE (dibucaine number also, to rule out genetic deficiencies). In view of the rapid reversibility, particular care should be taken to use analytical procedures that include immediate analysis and a careful control of sample dilutions (section 2.3). Chemical analysis of body fluids (gastric lavage, blood, urine) should be performed for the identification of the chemical(s).

9.3.1 Biochemical methods for measurement of effects

AChE present in human erythrocytes is the same as the enzymes present in the target synapses. Measurements of AChE in RBC is therefore taken to mirror the effects in the target organs. However, it must be borne in mind that this assumption is only correct when the carbamate has equal access to blood and the synapses, and that AChE inhibition by carbamates is reversible.

In the case of acute poisoning, a high inhibition of RBC-AChE is pathognomonic, but, in the follow-up of the intoxication, it might not be correlated with the severity of symptoms. However, monitoring of pre- and post-exposure levels of AChE in RBC gives a good measure of the effect of an exposure (Kaloyanova, 1976). In cases when the pre-exposure AChE level is not known (as in accidental poisoning), reference can be made to a mean population AChE activity. Blood-plasma contains a related enzyme called ChE or pseudo-ChE, which contributes to the whole blood enzymatic activity; the extent of the contribution of plasma-ChE will depend on which substrate was used and at what concentration. PseudoChE has no known physiological function, but can be inhibited selectively by some carbamates without causing a toxic response.

The sensitivities of AChE and pseudoChE to inhibitors differ so that measurements of the ability of whole blood samples to hydrolyse the usual analytical substrates give only an approximate estimate of the activity of the erythrocyte-AChE. However, under many field conditions, procedures using whole blood are more practical than those using separated erythrocytes.

9.4 Treatment of Acute Poisoning by Carbamate Insecticides

All cases of carbamate poisoning should be dealt with as an emergency and the patient should be hospitalized as quickly as possible.

Extensive descriptions of the treatment of poisoning by anticholinesterase agents are given in several major references (Kagan, 1977; Taylor, 1980; Plestina, 1984).

The treatment is based on:

(a) minimizing the absorption;

(b) general supportive treatment; and

(c) specific pharmacological treatment.

9.4.1 Minimizing the absorption

When dermal exposure occurs, decontamination procedures include removal of contaminated clothes and washing of the skin with alkaline soap or with a sodium bicarbonate solution. Particular care should be taken in the cleaning of the skin area where venupuncture is performed. Blood might be contaminated with carbamates and therefore inaccurate measures of ChE inhibition might result. Extensive eye irrigation with water or saline should also be performed. In the case of ingestion, vomiting can be induced, if the patient is conscious, by the administration of ipecacuanha syrup (10 - 30 ml) followed by 200 ml of water. However, this treatment is contraindicated in the case of pesticides dissolved in hydrocarbon solvents. Gastric lavage (with the addition of bicarbonate solution or activated charcoal) can also be performed, particularly in unconscious patients, taking care to prevent aspiration of fluids into the lungs (i.e., only after a tracheal tube has been put in place).

The volumes of the fluids introduced in the stomach should be recorded and samples of gastric lavage frozen and stored for subsequent chemical analysis. If the formulation of the pesticide involved is available, it should also be stored for further analysis (i.e., detection of toxicologically relevant impurities). A purge to remove the ingested compound can be administered.

9.4.2 General supportive treatment

Artificial respiration (via a tracheal tube) should be started at the first sign of respiratory failure and maintained for as long as necessary.

Cautious administration of fluids is advised as well as general supportive and symptomatic pharmacological treatment and absolute rest.

9.4.3 Specific pharmacological treatment

9.4.3.1 Atropine

Atropine should be given, beginning with 2 mg iv repeated at 15 to 30-min intervals. The dose and the frequency of atropine treatment varies from çase to case, but should maintain the patient fully atropinized (dilated pupils, dry mouth, skin flushing, etc.).

9.4.3.2 Oxime reactivators

There is no rational basis for using these drugs. Furthermore, some unconfirmed reports suggest an increased toxicity of carbamates when oximes have been administered.

9.4.3.3 Diazepam

Diazepam should be included in the therapy of all but the mildest cases. Besides relieving anxiety it appears to counteract some aspects of CNS-derived symptoms that are not affected by atropine. Doses of 10 mg sc or iv are appropriate and may be repeated as required.

Other centrally acting drugs and drugs that may depress respiration are not usually recommended in the absence of artificial respiration procedures.

10. PREVIOUS EVALUATIONS BY INTERNATIONAL BODIES

For more than 20 years, the Joint FAO/WHO Meetings on Pesticide Residues (JMPR) and the International Agency for Research on Cancer (IARC) have been evaluating the toxicity and carcinogenicity data on the different carbamates. As part of the JMPR meeting, the FAO deals with the use and application of these pesticides and proposes residue limits. Data and conclusions of the meetings were initially published in the WHO Technical Report Series but, since 1977, they have been published regularly in the FAO Plant Production and Protection Papers.

It is not possible to review completely the results of the discussions on the approximately 50 carbamates mentioned in these documents. However, Annex III contains a list of the JMPR meetings in which carbamates have been evaluated together with appropriate references. Other information available in Annex III indicates whether IARC evaluations and WHO/FAO Data Sheets are available, and whether an IRPTC data profile and legal file exist.

REFERENCES

ABDEL-AAL, Y.A.I. & FAHMY, M.A.H. (1977) Molecular modification of carbamates in relation to their selective toxicity. In: Proceedings of the British Crop Protection Conference on Pests and Diseases, Brighton, Nov. 1977, British Crop Protection Council, Croydon, pp. 155-159.

ABD-ELROAF, T.K., FAHMY, M.A.H., FUKUTO, T.R., & EL-SEBAE, A.H. (1977) Anticholinesterase activity and toxicity of substituted phenyl methylcarbamates to honey bee. J. econ. Entomol., 70(1): 78-82.

ABDEL-WAHAB, A.M., KUHR, R.J., & CASIDA, J.E. (1966) Fate of ^{14}C-carbonyl-labelled aryl methylcarbamate insecticide chemicals in and on bean plants. J. agric. food Chem., 14: 290-298.

ADDISON, J.B., SILK, P.J., & UNGER, I. (1973) The photochemical reactions of carbamates. III. The solution photochemistry of metacrate, 3-methylphenyl-N-methyl-carbamate. Int. J. environ. anal. Chem., 3: 73-79.

ADDISON, J.B., SILK, P.J., & UNGER, I. (1974) The solution photochemistry of metacil (4-dimethyl-amino-m-tolyl-N-methyl carbamate) and Landrin (3,4,5-trimethylphenyl-N-methyl-carbamate). Bull. environ. Contam. Toxicol., 11: 250-255.

AHDAYA, S.M., MONROE, R.J., & GUTHRIE, F.E. (1981) Absorption and distribution of intubated insecticides in fasted mice. Pestic. Biochem. Physiol., 16(1): 38-46.

ALDRIDGE, W.N. & MAGOS, L. (1978) Carbamates, thiocarbamates, and dithiocarbamates, Luxembourg, Commission of the European Communities.

ALDRIDGE, W.N. & REINER, E. (1972) Enzyme inhibitors as substrates. Interaction of esterases with esters of organophosphorus and carbamic acids, Amsterdam, Oxford, New York, Elsevier Science Publishers, 328 pp.

ALY, O.M. & EL-DIB, M.A. (1971) Photodecomposition of some carbamate insecticides in aquatic environments. In: Faust, S.D. & Hunter, J.V., ed. Organic compounds in aquatic environments, New York, Marcel Dekker, pp. 469-493.

ALY, O.M. & EL-DIB, M.A. (1972) Studies of the persistence of some carbamate insecticides in the aquatic environment. In:

Fate of organic compounds in the aquatic environment. III. Advances in chemistry series, Washington DC, American Chemical Society, pp. 210-243.

ANDERSON, C.A. (1973) Baygon. In: Sherma, J. & Zweig, G., ed. Analytical methods for pesticides and plant growth regulators, New York, London, Academic Press, Vol. 7, pp. 163-178.

ANDERSON, K.J., LEIGHTY, E.G., & TAKAHASHI, M.T. (1972) Evaluation of herbicides for possible mutagenic properties. J. agric. food Chem., 20: 649-656.

ANDRIANOVA, M.M. & ALEKSEEV, I.V. (1970) [On the carcinogenous properties of the pesticides sevine, maneb, ciram and cineb.] Vopr. Pitan., 29(6): 71-74 (in Russian, with English summary).

ARMSTRONG, D.A. & MILLEMANN, R.E. (1974a) Effects of the insecticide carbaryl on clams and some other intertidal mud flat animals. J. Fish. Res. Board Can., 31: 466-470.

ARMSTRONG, D.A. & MILLEMANN, R.E. (1974b) Effects of the insecticide Sevin and its first hydrolytic product 1-naphthol on some early developmental stages of the bay mussel Mytilus edulis. Mar. Biol., 28: 11-15.

ASHTON, F.M. & CRAFTS, A.S. (1973) Carbamates. In: Mode of action of herbicides, New York, Wiley Interscience, pp. 200-220.

ATKINS, E.L., GREYWOOD, E.A., & MCDONALD, R.L. (1973) Toxicity of pesticides and other agricultural chemicals to honey bees. Laboratory studies, Riverside, USA, University of California, Agriculture Extension Service.

AUGUSTINSSON, K.B. & CASIDA, J.E. (1959) Enzymic hydrolysis of N,N-dimethylcarbamoyl fluoride. Biochem. Pharmacol., 3: 60-67.

AUSTIN D.J. & BRIGGS, G.G. (1976) A new extraction method for benomyl residues in soil and its application in movement and persistence studies. Pestic. Sci., 7: 201-210.

AUSTIN, D.J., LORD, K.A., & WILLIAMS, I.H. (1976) High pressure liquid chromatography of benzimidazoles. Pestic. Sci., 7: 211-222.

BAKER, P.B. & HOODLESS, R.A. (1974) Analytical methods for the detection and determination of residues of systemic fungicides. Pestic. Sci., 5: 465-472.

BALDWIN, R.E., FREED, V.H., & FANG, S.C. (1954) Absorption and translocation of carbon-14-applied as O-isopropyl N-phenylcarbamate in Avena and Zea. J. agric. food Chem., 2(8): 428-430.

BERAUD, M., PIPY, B., DERACHE, R., & GAILLARD, D. (1979) Formation d'un cancérigène le N-nitrosocarbaryl par interactions entre un insecticide de la série des carbamates le carbaryl et le nitrite de sodium dans le suc gastrique de rat. Food Cosmet. Toxicol., 17: 579-583.

BIGNAMI, M., AULICINO, F., VELCICH, A., CARERE, A., & MORPURGO, G. (1977) Mutagenic and recombinogenic action of pesticides in Aspergillus nidulans. Mutat. Res., 46: 395-402.

BLACK, A.L., CHIU, Y.C., FAHMY, M.A.H., & FUKUTO, T.R. (1973) Selective toxicity of N-sulfenylated derivatives of insecticidal methylcarbamate esters. J. agric. food Chem., 21: 743-747.

BLEIDNER, W.E., MORALES, R., & HOLT, R.F. (1978) Benomyl. In: Zweig, G. & Sherma, J., ed. Analytical methods for pesticides and plant growth regulators, New York, London, Academic Press, pp. 157-171.

BLEVINS, R.D., LYINSKY, W., & REGAN, J.D. (1977) Nitrosated methylcarbamate insecticides. Effect on the DNA of human cells. Mutat. Res., 44: 1-7.

BOBIK, A., HOLDER, G.M., & RYAN, A.J. (1972) Excretory and metabolic studies of isopropyl N-(3-chlorophenyl)carbamate in the rat. Food Cosmet. Toxicol., 10: 163-170.

BOMIRSKA, T. & WINIARSKA, A. (1972) [Toxicology of carbamates in the light of intoxication with Baygon insecticide.] Pol. Tyg. Lek., 27: 1448-1450 (in Polish).

BORZSONYI, M. & CSIK, M. (1975) Induction of malignant lymphomas in Swiss mice by N-nitroso compounds formed in vivo. Int. J. Cancer, 15: 830-838.

BORZSONYI, M. & PINTER, A. (1977) The carcinogenicity of N-nitroso compounds formed endogenously in mice from benzimidazole carbamate pesticides. Neoplasma, 24: 119-122.

BOWMAN, M.C. & BEROZA, M. (1969) Determination of Mesurol and five of its metabolites in apples, pears and corn by gas chromatography. J. Assoc. Off. Anal. Chem., 52: 1054-1063.

BROCKELSBY, C.H. & MUGGLETON, D.F. (1973) Asulam. In: Sherma, J. & Zweig, G., ed. Analytical methods for pesticides and plant growth regulators, New York, London, Academic Press, Vol. 7, pp. 497-508.

CARERE, A., ORTALI, V.A., CARDAMONE, G., TORRACCA, A.M., & RASCHETTI, R. (1978) Microbiological mutagenicity studies of pesticides in vitro. Mutat. Res., 57: 277-286.

CARPENTER, C.P., WEIL, C.S., PALM, P.E., WOODSIDE, M.W., NAIR, J.H., & SMYTH, H.F. (1961) Mammalian toxicity of 1-naphthyl N-methylcarbamate (Sevin insecticide). J. agric. food Chem., 9: 30-39.

CARTER, S.D. & LASKEY, J.W. (1982) Effect of benomyl on reproduction in the male rat. Toxicol. Lett., 11: 87-94.

CASIDA, J.E. & AUGUSTINSSON, K.B. (1959) Reaction of plasma albumin with 1-naphthyl N-methylcarbamate and certain other esters. Biochem. Biophys. Acta, 36: 411-426.

CHIBA, M. & DOORNBOS, F. (1974) Instability of benomyl in various conditions. Bull. environ. Contam. Toxicol., 11: 273-274.

CHIN, W.T., DUANE, W.C., SZALKOWSKI, M.B., & STALLARD, D.E. (1975) Gas-chromatographic determination of thiofanox residues in soil, plants, and water. J. agric. food Chem., 23(5): 963-966.

CLARK, C.G. & WRIGHT, S.J.L. (1970) Degradation of the herbicide isopropyl N-phenylcarbamate by Arthrobacter and Achromobacter sp. from soil. Soil Biol. Biochem., 2: 217.

COLLINS, T.F.X., HANSEN, W.H., & KEELER, H.V. (1971) The effect of carbaryl (Sevin) on reproduction of the rat and the gerbil. Toxicol. appl. Pharmacol., 19: 202-216.

CROSBY, D.G., LEITIS, E., & WINTERLIN, W.L. (1965) Photodecomposition of carbamate insecticides. J. agric. food Chem., 13(3): 204-207.

DASSENOY, B. & MEYER, J.A. (1973) Mutagenic effect of benomyl on Fusarium oxysporum. Mutat. Res., 21: 119-120.

DAWSON, J.A., HEATH, D.F., ROSE, J.A., THAIN, E.M., & WARD, J.B. (1964) The excretion by humans of the phenol derived in vivo from 2-isopropoxyphenyl-N-methylcarbamate. Bull. World Health Org., 30: 127-134.

DELATOUR, P. & RICHARD, Y. (1976) Propriétés embryotoxiques et antimitotiques en série benzimidazole. Therapie, 31(4): 505-515.

DE ROSE, H.R. (1951) Crabgrass inhibition with O-isopropyl-N-(3-chlorophenyl)carbamate. Agron. J., 43: 139-142.

DIKSHITH, T.S.S., GUPTA, P.K., GAUR, J.S., DATTA, K.K., & MATHUR, A.K. (1976) Ninety-day toxicity of carbaryl in male rats. Environ. Res., 12: 161-170.

DOROUGH, H.W. & CASIDA, J.E. (1964) Nature of certain carbamate metabolites of the insecticide Sevin. J. agric. food Chem., 12(4): 294-304.

DOUCH, P.G.C. (1973) The metabolism of benomyl fungicide in animals. Xenobiotica, 3: 367-380.

EBERLE, D.O. & GUNTHER, F.A. (1965) Chromatographic, spectrometric, and irradiation behaviour of five carbamate insecticides. J. Assoc. Off. Agric. Chem., 48:(5) 927-937.

EDWARDS, C.A. & THOMPSON, A.R. (1973) Pesticides and the soil fauna. Residue Rev., 45: 1-79.

EISENBRAND, G., UNGERER, O., & PREUSSMANN, R. (1974) Rapid formation of carcinogenic N-nitrosamines by interaction of nitrite with fungicides derived from dithiocarbamic acid in vitro under stimulated gastric conditions and in vivo in the rat stomach. Food Cosmet. Toxicol., 12: 229-232.

EISENBRAND, G., IVANKOVIC, S., PREUSSMANN, R., SCHMAHL, D., & WIESSLER, M. (1975a) Some recent results on the chemistry, formation and biological activity of N-nitroso compounds. Gaun Monogr. Cancer Res., 17: 133-143.

EISENBRAND, G., UNGERER, O., & PREUSSMANN, R. (1975b) The reaction of nitrite with pesticides. II. Formation, chemical properties, and carcinogenic activity of the N-nitroso derivative of N-methyl-1-naphthylcarbamate (carbaryl). Food Cosmet. Toxicol., 13: 365-367.

ELESPURU, R.K. & LIJINSKY, W. (1973) The formation of carcinogenic nitroso compounds from nitrite and some types of agricultural chemicals. Food Cosmet. Toxicol., 11: 807-817.

ELESPURU, R.K., LIJINSKY, W., & SETLOW, J.K. (1974) Nitrosocarbaryl as a potent mutagen of environmental significance. Nature (Lond.), 247: 386-387.

ELLMAN, G.L., COURTNEY, K.D., ANDRES, V. Jr, & FEATHERSTONE, R.M. (1961) A new and rapid colorimetric determination of acetylcholinesterase activity. Biochem. Pharmacol., 7: 88-95.

FANG, S.C., FALLIN, E., MONTGOMERY, M.L., & FREED, V.H. (1974) Metabolic studies of ^{14}C-labelled propham and chlorpropham in the female rat. Pestic. Biochem. Physiol., 4: 1-11.

FAHRIG, R. & SEILER, J.P. (1979) Dose and effect of methyl-2-benzimidazolylcarbamate in the "mammalian spot test", an in vivo method for the detection of genetic alterations in somatic cells of mice. Chem.-biol. Interact., 26: 115-120.

FAO (1985) Production yearbook, 1984, Rome, Food and Agriculture Organization of the United Nations.

FAO/WHO (1964) Evaluation of the toxicity of pesticide residues in food. Report of a Joint Meeting of the FAO Committee on Pesticides in Agriculture and the WHO Expert Committee on Pesticide Residues, Geneva, World Health Organization (FAO Meeting Report No. PL:1963/13; WHO/Food Add./23.64).

FAO/WHO (1965a) Evaluation of the toxicity of pesticide residues in food. Report of the Second Joint Meeting of the FAO Committee on Pesticides in Agriculture and the WHO Expert Committee on Pesticide Residues, Geneva, World Health Organization (FAO Meeting Report No. PL:1965/10; WHO Food Add./26.65).

FAO/WHO (1965b) Evaluation of the toxicity of pesticide residues in food, Geneva, World Health Organization (FAO Meeting Report No. PL:1965/10/1; WHO Food Add./27.65).

FAO/WHO (1967a) Pesticide residues in food. Joint report of the FAO Working Party on Pesticide Residues and the WHO Expert Committee on Pesticide Residues, Geneva, World Health Organization (FAO Agricultural Studies No. 73; WHO Technical Report Series No. 370).

FAO/WHO (1967b) *Evaluation of some pesticide residues in food*, Geneva, World Health Organization (FAO: PL/CP/15; WHO Food Add./67.32).

FAO/WHO (1968a) *Pesticide residues in food. Report of the 1967 Joint Meeting of the FAO Working Party of Experts on Pesticide Residues and the WHO Expert Committee on Pesticide Residues*, Geneva, World Health Organization (FAO Meeting Report No. PL:1967/M/11; WHO Technical Report Series No. 391.

FAO/WHO (1968b) *1967 Evaluation of some pesticide residues in food*, Geneva, World Health Organization (FAO PL:1967/M/11/1; WHO Food Add./68.30).

FAO/WHO (1969a) *Pesticide residues in food. Report of the 1968 Joint Meeting of the FAO Working Party of Experts on Pesticide Residues and the WHO Expert Committee on Pesticide Residues*, Geneva, World Health Organization (FAO Agricultural Studies No. 78; WHO Technical Report Series No. 417).

FAO/WHO (1969b) *1968 Evaluation of some pesticide residues in food*, Geneva, World Health Organization (FAO PL:1968/M/9/1; WHO Food Add./69.35).

FAO/WHO (1970a) *Pesticide residues in food. Report of the 1969 Joint Meeting of the FAO Working Party of Experts on Pesticide Residues and the WHO Expert Group on Pesticide Residues*, Geneva, World Health Organization (FAO Agricultural Studies No. 84; WHO Technical Report Series No. 458).

FAO/WHO (1970b) *1969 Evaluation of some pesticide residues in food*, Geneva, World Health Organization (FAO PL:1969/M/17/1; WHO Food Add./70.38).

FAO/WHO (1971a) *Pesticide residues in food. Report of the 1970 Joint Meeting of the FAO Working Party of Experts on Pesticide Residues and the WHO Expert Committee on Pesticide Residues*, Geneva, World Health Organization (FAO Agricultural Studies No. 87; WHO Technical Report Series No. 474).

FAO/WHO (1971b) *1970 Evaluation of some pesticide residues in food*, Geneva, World Health Organization (AGP 1979/M/12/1; WHO Food Add./71.42).

FAO/WHO (1972a) *Pesticide residues in food. Report of the 1971 Joint Meeting of the FAO Working Party of Experts on Pesticide Residues and the WHO Expert Committee on Pesticide Residues*, Geneva, World Health Organization (FAO Agricultural Studies No. 88; WHO Technical Report Series No. 502).

FAO/WHO (1974a) Pesticide residues in food. Report of the 1973 Joint Meeting of the FAO Working Party of Experts on Pesticide Residues and the WHO Expert Committee on Pesticide Residues, Geneva, World Health Organization (FAO Agricultural Studies No. 92; WHO Technical Report Series No. 545).

FAO/WHO (1974b) 1973 Evaluation of some pesticide residues in food, Geneva, World Health Organization (AGP 1973/M/9/1; WHO Pesticide Residues Series No. 3).

FAO/WHO (1975a) Pesticide residues in food. Report of the 1974 Joint Meeting of the FAO Working Party of Experts on Pesticide Residues and the WHO Expert Committee on Pesticide Residues, Geneva, World Health Organization (FAO Agricultural Studies No. 97; WHO Technical Report Series No. 574).

FAO/WHO (1975b) 1974 Evaluation of some pesticide residues in food, Geneva, World Health Organization (AGP 1974/M/11; WHO Pesticide Residues Series No. 4).

FAO/WHO (1976a) Pesticide residues in food. Report of the 1975 Joint Meeting of the FAO Working Party of Experts on Pesticide Residues and the WHO Expert Committee on Pesticide Residues, Geneva, World Health Organization (FAO Plant Production and Protection Series No. 1; WHO Technical Report Series No. 592).

FAO/WHO (1976b) 1975 Evaluation of some pesticide residues in food, Geneva, World Health Organization (AGP 1975/M/13; WHO Pesticide Residues Series No. 5).

FAO/WHO (1977a) Pesticide residues in food. Report of the 1976 Joint Meeting of the FAO Panel of Experts on Pesticide Residues and the Environment and the WHO Expert Group on Pesticide Residues, Geneva, World Health Organization (FAO Food and Nutrition Series No. 9; FAO Plant Production and Protection Series No. 8; WHO Technical Report Series No. 612).

FAO/WHO (1977b) 1976 Evaluation of some pesticide residues in food, Rome, Food and Agriculture Organization of the United Nations (AGP 1977/M/14).

FAO/WHO (1978a) Pesticide residues in food. Report of the 1977 Joint Meeting of the FAO Panel of Experts on Pesticide Residues and the Environment and the WHO Expert Group on Pesticide Residues, Rome, Food and Agriculture Organization of the United Nations (FAO Plant Production and Protection Paper 10 Rev).

FAO/WHO (1978b) 1977 Evaluation of some pesticide residues in food, Rome, Food and Agriculture Organization of the United Nations (FAO Plant Production and Protection Paper 10 Suppl.).

FAO/WHO (1979a) Pesticide residues in food. Report of the 1978 Joint Meeting of the FAO Panel of Experts on Pesticide Residues in Food and the Environment and the WHO Expert Group on Pesticide Residues, Rome, Food and Agriculture Organization of the United Nations (FAO Plant Production and Protection Paper 15).

FAO/WHO (1979b) 1978 Evaluation of some pesticide residues in food, Rome, Food and Agriculture Organization of the United Nations (FAO Plant Production and Protection Paper 15 Suppl.).

FAO/WHO (1980a) Pesticide residues in food. Report of the 1979 Joint Meeting of the FAO Panel of Experts on Pesticide Residues in Food and the Environment and the WHO Expert Group on Pesticide Residues, Rome, Food and Agriculture Organization of the United Nations (FAO Plant Production and Protection Paper 20).

FAO/WHO (1980b) 1979 Evaluations of some pesticide residues in food, Rome, Food and Agriculture Organization of the United Nations (FAO Plant Production and Protection Paper 20 Suppl.).

FAO/WHO (1981a) Pesticide residues in food. Report of the 1980 Joint Meeting of the FAO Panel of Experts on Pesticide Residues in Food and the Environment and the WHO Expert Group on Pesticide Residues, Rome, Food and Agriculture Organization of the United Nations (FAO Plant Production and Protection Paper 26).

FAO/WHO (1981b) 1980 Evaluation of some pesticide residues in food, Rome, Food and Agriculture Organization of the United Nations (FAO Plant Production and Protection Paper 26 Suppl.).

FAO/WHO (1982a) Pesticide residues in food. Report of the 1981 Joint Meeting of the FAO Panel of Experts on Pesticide Residues in Food and the Environment and the WHO Expert Group on Pesticide Residues, Rome, Food and Agriculture Organization of the United Nations (FAO Plant Production and Protection Paper 37).

FAO/WHO (1982b) 1981 Evaluation of some pesticide residues in food, Rome, Food and Agriculture Organization of the United Nations (FAO Plant Production and Protection Paper 42).

FAO/WHO (1983a) Pesticide residues in food. Report of the 1982 Joint Meeting of the FAO Panel of Experts on Pesticide Residues in Food and the Environment and the WHO Expert Group on Pesticide Residues, Rome, Food and Agriculture Organization of the United Nations (FAO Plant Production and Protection Paper 46).

FAO/WHO (1983b) 1982 Evaluation of some pesticide residues in food, Rome, Food and Agriculture Organization of the United Nations (FAO Plant Production and Protection Paper 49).

FAO/WHO (1984) Pesticide residues in food. Report of the 1983 Joint Meeting of the FAO Panel of Experts on Pesticide Residues in Food and the Environment and the WHO Expert Group on Pesticide Residues, Rome, Food and Agriculture Organization of the United Nations (FAO Plant Production and Protection Paper 56).

FAO/WHO (1985a) 1983 Evaluation of some pesticide residues in food, Rome, Food and Agriculture Organization of the United Nations.

FAO/WHO (1985b) Pesticide residues in food. Report of the 1984 Joint Meeting of the FAO Working Party of Experts on Pesticide Residues and the WHO Expert Group on Pesticide Residues, Rome, Food and Agriculture Organization of the United Nations (FAO Plant Production and Protection Paper 62).

FAO/WHO (1985c) 1984 Evaluation of some pesticide residues in food, Rome, Food and Agriculture Organization of the United Nations.

FAO/WHO (1986) Pesticide residues in food. Report of the 1985 Joint Meeting of the FAO Working Party of Experts on Pesticide Residues and the WHO Expert Group on Pesticide Residues, Rome, Food and Agriculture Organization of the United Nations (FAO Plant Production and Protection Paper 68).

FAO/WHO (in press) 1985 Evaluation of some pesticide residues in food, Rome, Food and Agriculture Organization of the United Nations.

FARAGO, A. (1969) [Fatal suicidal poisoning with Sevin (1-naphthyl-N-methyl carbamate).] Arch. Toxikol., 24: 309-315 (in German).

FICSOR, G., BORDAS, S., & STEWART, S.J. (1978) Mutagenicity testing of benomyl methyl-2-benzimidazole carbamate, streptozotocin, and N-methyl-N'-nitro-N'-nitrosoguanidine in

Salmonella typhimurium in vitro and in rodent host-mediated assays. Mutat. Res., 51: 151-164.

FINNISH NATIONAL BOARD OF HEALTH TOXICOLOGY GROUP (1982) The toxicity of the benzimidazol type fungicides: benomyl, carbendazim, and thiophanate-methyl (second report), Helsinki, Finnish National Board of Health.

FILIP, Z. (1974) [Pesticides in powder form.] Vesmir, 53: 243-244 (in Czech).

FUKUTO, T.R. (1972) Metabolism of carbamate insecticides. Drug Metab. Rev., 1: 117-150.

GARDINER, J.A., KIRKLAND, J.J., KLOPPING, H.L., & SHERMAN, H. (1974) Fate of benomyl in animals. J. agric. food Chem., 22(3): 419-427.

GHADIRI M. & GREENWOOD, D.A. (1966) Toxicity and biologic effects of malathion, phosdrin, and sevin in the chick embryo. Toxicol. appl. Pharmacol., 8: 342.

GHADIRI M., GREENWOOD, D.A., & BINNS, W. (1967) Feeding of malathion and carbaryl to laying hens and roosters. Toxicol. appl. Pharmacol., 10: 392.

GOLD, R.E., LEAVITT, J.R.C., HOLCSLAW, T., & TUPY, D. (1982) Exposure of urban applications to carbaryl. Arch. environ. Contam. Toxicol., 11: 63-67.

GOLDBERG, M.E., JOHNSON, H.E., KNAAK, J.B., & SMYTH, H.F., Jr (1963) Psychopharmacological effects of reversible cholinesterase inhibition induced by N-methyl-3-isopropylphenyl carbamate (compound 10854). J. Pharm. exp. Ther., 141: 244-252.

GRAY, R.A. (1971) Behaviour, persistence, and degradation of carbamate and thiocarbamate herbicides in the environment. In: Proceedings of the California Weed Control Conference, 18-20 January 1971, Mountain View, California, Stauffer Western Research Centre, pp. 128-143.

GUARDIGLI, A., CHOW, W., & LEFAR, M.S. (1977) Determination of carbetamid residues and its aniline metabolite. J. agric. food Chem., 20: 348-350.

GUPTA, A.K. & LEGATOR, M.S. (1975) Chromosome aberrations in cultured human leukocytes after treatment with the fungicide benlate. In: Proceedings of a Symposium on Mutagenic

Carcinogenic and Teratogenic Chemistry, New Delhi, India, Department of Atomic Energy, pp. 95-103.

HANSEN, D.J. (1969) Avoidance of pesticides by untrained sheepshead minnows. Trans. Am. Fish. Soc., 98: 426-429.

HARVEY, J., Jr & HAN, J.C.Y. (1978a) Metabolism of oxamyl and selected metabolites in the rat. J. agric. food Chem., 26(4): 902-910.

HARVEY, J., Jr & HAN, J.C.Y. (1978b) Decomposition of oxamyl in soil and water. J. agric. food Chem., 26(3): 536-541.

HARVEY, J., Jr & PEASE, H.L. (1973) Decomposition of methomyl in soil. J. agric. food Chem., 21(5): 784-786.

HARVEY, J., Jr & REISER, R.W. (1973) Metabolism of methomyl in tobacco, corn, and cabbage. J. agric. food Chem., 21(5): 775-783.

HARVEY, J., Jr, JELINEK, A.G., & SHERMAN, H. (1973) Metabolism of methomyl in the rat. J. agric. food Chem., 21(5): 769-775.

HARVEY, J., Jr, HAN, J.C.Y., & REISER, R.W. (1978) Metabolism of oxamyl in plants. J. agric. food Chem., 26(3): 529-536.

HEYBROEK, W.M.H., MUGGLETON, D.F., & PARKE, D.V. (1984) Metabolism of the carbamate herbicide asulam in the rat. Xenobiotica, 14:(3) 235-247.

IARC (1976) Some carbamates, thiocarbarmates, and carbazides, Lyons, International Agency for Research on Cancer (IARC Monographs on the Evaluation of Carcinogenic Risk of Chemicals to Man, Vol. 12).

IARC (1983) Miscellaneous pesticides, Lyons, International Agency for Research on Cancer (IARC Monographs on the Evaluation of Carcinogenic Risk of Chemicals to Man, Vol. 30; Appendix 359-406).

INNES, J.R.M., ULLAND, B.M., VALLERIO, M.G., PETRUCELLI, L., FISHBEIN, L., HART, E.R., PALLOTTA, A.J., BATES, R.R., FALK, H.L., GART, J.J., KLEIN, M., MITCHELL, J., & PETERS, J. (1969) Bioassay of pesticides and industrial chemicals for tumorigenicity in mice: a preliminary note. J. Natl Cancer Inst., 42: 1101-1114.

IRPTC (1983) IRPTC legal file 1983, Geneva, International Register of Potentially Toxic Chemicals, United Nations Environment Programme.

IVANOVA-CHEMISHANSKA, L. & ANTOV, G. (1976) Changes in the activity of some enzymes under the action of benomyl. Hig. i Zdraveopazvane, Sofia, 19(5): 431-435 (in Bulgarian, with English summary).

IVANOVA-CHEMISHANSKA, L. & ANTOV, G. (1980) Changes in the gonads, reproduction, and generations of white rats under the effect of some pesticides. Arch. Toxicol., 4(Suppl.): 459-462.

IVANOVA-CHEMISHANSKA, L., ANTOV, G., MIHAILOVA, A., & HINKOVA, L. (1980) Study on the toxicology of benomyl. In: The Fourth International Congress of Pesticide Chemistry, Zurich, Switzerland, 24-28 July, 1978, pp. 11-559.

IZMIROVA, N. (1980) Methods for determination of exposure of agricultural workers to organophosphorous pesticides. In: Tordoir, W.F. & van Heemstra-Lequin, E.A.H., ed. Field worker exposure during pesticide application, Amsterdam, Oxford, New York, Elsevier Scientific Publishers, pp. 169-172 (Studies in Environmental Science 7).

IZMIROVA, N., MILCEVA, V., MONOV, A., & KALOYANOVA, F. (1981) [Acute carbofuran intoxication.] Hig. i zdraveopazvane, 24(5): 445-448 (in Bulgarian, with English summary).

JACKSON, J.A., CHART, I.S., SANDERSON, J.H., & GARNER, R. (1977) Pirimicarb induced immune haemolytic anaemia in dogs. Scand. J. Haematol., 19: 360-366.

JOHNSON, M.K. (1970) Organophosphorus and other inhibitors of brain "neurotoxic esterase" and the development of delayed neurotoxicity in hens. Biochem. J., 120: 523-531.

JOHNSON, W.W. & FINLEY, M.T. (1980) Handbook of acute toxicity of chemicals to fish and aquatic invertebrates, Washington DC, US Fish and Wildlife Service (Resource Publication No. 137).

KAGAN, J.S. (1977) [The toxicity of organophosphorus pesticides,] Moscow (in Russian).

KALOYANOVA, F. (1976) Cholinesterase activity as a biochemical indicator for monitoring exposure to certain pesticides. In: Proceedings of the International Symposium on

Environmental Monitoring, 14-19 September, 1975, Las Vegas, Nevada, Vol. 1, pp. 1-4, New York, Institute of Electrical & Electronics Engineers, Inc.

KANAZAWA, J. (1975) Uptake and excretion of organophosphorus and carbamate insecticides by fresh-water fish Motsugo pseudoras bora parva. Bull. environ. Contam. Toxicol., 14: 346-352.

KAPLAN, A.M. & SHERMAN, H. (1977) Toxicity studies with methyl-N-methylamino-carbonyl-oxyethanimidothioate. Toxicol. appl. Pharmacol., 40: 1-17.

KAPPAS, A., GEORGOPOULOS, S.G., & HASTIE, A.C. (1974) On the genetic activity of benzimidazole and thiophanate fungicides on diploid Aspergillus nidulans. Mutat. Res., 26: 17-27.

KAPPAS, A., GREEN, M.H.L., BRIDGES, B.A., ROGERS, A.M., & MURIEL, W.J. (1976) Benomyl: a novel type of base analogue mutagen. Mutat. Res., 40: 379-382.

KAVLOCK, R.J., CHERNOFF, N., GRAY, L.E., Jr, GRAY, J.A., & WHITEHOUSE, D. (1982) Teratogenic effects of benomyl in the Wistar rat and CD-1 mouse, with emphasis on the route of administration. Toxicol. appl. Pharmacol., 62: 44-54.

KAUFMAN, D.D. & BLAKE, J. (1973) Microbial degradation of several acetamide, acrylamide, carbamate, toluidine, and urea pesticides. Soil Biol. Biochem., 5: 297-308.

KAUFMAN, D.D. & KEARNEY, P.C. (1965) Microbial degradation of isopropyl-N-(3-chlorophenyl carbamate and 2-chloroethyl-N-(3-chlorophenyl carbamate. Appl. Microbiol., 13(3): 443-446.

KHERA, K.S. (1966) Toxic and teratogenic effects of insecticides in duck and chick embryos. Toxicol. appl. Pharmacol., 8: 345.

KIMMERLE, G. & IYATOMI, A. (1976) Toxicity of propoxur to rats by subacute inhalation. Jpn J. ind. Health, 18: 375-382.

KNAAK, J.B., TALLANT, M.J., BARTLEY, W.Y., & SULLIVAN, L.J. (1965) The metabolism of carbaryl in the rat, guinea pig and man. J. agric. food Chem., 13(4): 537-543.

KNAAK, J.B., TALLANT, M.J., & SULLIVAN, L.J. (1966) The metabolism of 2-methyl-2-(methylthio)propionaldehyde-O-methylcarbamoyl)oxime in the rat. J. agric. food Chem., 14(6): 573-578.

KNAAK, J.B., TALLANT, M.J., KOZBELT, S.J., & SULLIVAN, L.J. (1968) Metabolism of carbaryl in man, monkey, pig, and sheep. J. agric. food Chem., 16(3): 465-470.

KNOWLES, C.O. & SONAWANE, B.R. (1972) Ethyl m-hydroxy-carbanilate carbanilate (EP-475) metabolism in sugar beets. Bull. environ. Contam. Toxicol., 8: 73-76.

KOSSMANN, K. & JENNY, N.A. (1973) Phenmedipham. In: Sherma, J. & Zweig, G., ed. Analytical methods for pesticides and plant growth regulators, New York, London, Academic Press, Vol. 7, pp. 611-623.

KRECHNIAK, J. & FOSS, W. (1982) Cholinesterase activity in rats treated with propoxur. Bull. environ. Contam. Toxicol., 29(5): 599-604.

KUHR, R.J. (1968) Metabolism of methylcarbamate insecticide chemicals in plants. J. Sci. Food Agric. 44, (Suppl.): 43-49.

KWALICK, D.S. (1971) Physician's pesticides primer. J. Med. Soc. New Jersey, 68:(5) 375-377.

LAMB, M.J. & LILLY, L.J. (1980) An investigation of some genetic toxicological effects of the fungicide benomyl. Toxicology, 17: 83-95.

LEAVITT, J.R.C., GOLD, R.E., HOLCSLAW, T., & TUPY, D. (1982) Exposure of professional pesticide applicators to carbaryl. Arch. environ. Contam. Toxicol., 11: 57-62.

LEELING, N.C. & CASIDA, J.E. (1966) Metabolites of carbaryl (1-naphthyl methylcarmate) in mammals and enzymatic systems for their formation. J. agric. food Chem., 14: 281-290.

LEITCH, R.E. & PEASE, H.L. (1973) Lannate-methomyl. In: Sherma, J. & Zweig, G., ed. Analytical methods for pesticides and plant growth regulators, New York, London, Academic Press, Vol. 7, pp. 331-338.

LYINSKY, W. & TAYLOR, H.W. (1976) Carcinogenesis in Sprague Dawley rats of N-nitroso-N-alkylcarbamate esters. Cancer Lett., 1: 275-279.

MAKITA, T., HASHIMOTO, Y., & NOGUCHI, T. (1973) Mutagenic, cytogenetic, and teratogenic studies on thiophanate-methyl. Toxicol. appl. Pharmacol., 24: 206-215.

MARLIAC, J.P. (1964) Toxicity and teratogenic effects of 12 pesticides in the chick embryo. Fed. Proc. Fed. Am. Soc. Exp. Biol., 23: 105-126

MCCARTHY, J.F., FANCHER, O.E., KENNEDY, G.L., KEPLINGER, M.L., & CALANDRA, J.C. (1971) Reproduction and teratology studies with the insecticide carbofuran. Toxicol. appl. Pharmacology, 19: 370.

METCALF, R.L., FUKUTO, T.R., COLLINS, C., BORCK, K., BURK J., REYNOLDS, H.T., & OSMAN, M.F. (1966) Metabolism in plant and insect. Metabolism of 2-methyl-2-(methylthio)-propionaldehyde O-methylcarbamoyl-oxime in plant and insect. J. agric. food Chem., 14: 579-584.

MILLER, D.B. (1982) Neurotoxicity of the pesticidal carbamates. Neurobehav. Toxicol. Teratol., 4: 779-787.

MIRVISH, H.S.S. (1968) The carcinogenic action and metabolism of urethan and N-hydroxyurethan. Adv. Cancer Res., 11: 1-42.

MONTAZEMI, K. (1969) Toxicological studies of Baygon insecticide in Shabankareh area, Iran. Trop. geogr. Med., 21: 186-190.

MORSE, D.L. & BAKER, E.L., Jr (1979) Propanil-chloracne and methomyl toxicity in workers of a pesticide manufacturing plant. Clin. Tox., 15(1): 13-21.

MOSCIONI, A.D., ENGEL, J.L., & CASIDA, J.E. (1977) Kynurenine formamidase inhibition as a possible mechanism for certain teratogenic effect of organophosphorus and methylcarbamate insecticides in chicken embryos. Biochem. Pharmacol., 26: 2251-2258.

NAMBA, T., NOLTE, C.T., JACKREL, G., & GROB, D. (1971) Poisonning due to organophosphate insecticides: acute and chronic manifestations. Am. J. Med., 50: 475-492.

NATIONAL INSTITUTE OF HYGIENIC SCIENCES (1985) Letter to the International Programme on Chemical Safety, 1.10.85, Tokyo, Japan.

NISHIUCHI, Y. (1974) [Toxicity of pesticides formulations to some fresh-water organisms. XXIII.] Aquiculture, 22: 16-18 (in Japanese).

OGLE, R.E. & WARREN, G.F. (1954) Fate and activity of herbicides in soils. Weeds, 3: 257-273.

OLEFIR, A.I. & VINOGRADOVA, V.Kh. (1968) [A study of the embryotropic effect of the pesticides sevine and cyram.] Vrach. Delo., 11: 103-106 (in Russian).

OLIVER, J.E. (1981) Pesticide-derived nitrosoamines. Occurrence and environmental fate. In: Scanlan, R.A. & Tannenbaum, S.T., ed. N-nitroso compounds, Washington DC, American Chemical Society (ACS Symposium Series 174).

OONNITHAN, E.S. & CASIDA, J.E. (1968) Oxidation of methyl and dimethylcarbamate insecticide chemicals by microsomal enzymes and anticholinesterase activity of the metabolites. J. agric. food Chem., 16: 28-44.

ORLOVA, N.V. & ZHALBE, E.P. (1968) Experimental material contributive to the profile of maximal permissible amounts of Sevin in food products. Vopr. Pitan., 27: 49-55 (in Russian).

PAROCHETTI, J.V. & WARREN, G.F. (1966) Vapour losses of IPC and CIPC. Weeds, 14; 281-285.

PASTUSHENKO, T.B. (1983) [Study on the ganodotropic effects of methyl N-2-benzimidazol carbamate aiming at evaluating its effective concentrations.] Gig. i Sanit., 7: 83-85 (in Russian).

PAULSON, G.D., JACOBSEN, A.M., ZAYLSKIE, R.G., & FEIL, V.J. (1973) Isolation and identification of propham isopropylcarbanilate metabolites from the rat and goat. J. agric. food Chem., 21: 804-811.

PETROVA-VERGIEVA, T. (1979) [Changes in the intrauterine development of experimental animals under fundazol effect.] Hig. Zdraveopaz., 22: 487-495 (in Bulgarian).

PLESTINA, R. (1984) Prevention, diagnosis, and treatment of insecticide poisoning, Geneva, World Health Organization. (Unpublished report VBC/84.889).

PRENDEVILLE, G.N., ESHEL, Y., JAMES, C.S., WAREN, G.F.E., & SCHREIBER, M.M. (1968) Movement and metabolism of CIPC in resistent and susceptible species. Weed Sci., 16: 432-435.

PROCTOR, N.H., MOSCIONI, A.D., & CASIDA, J.E. (1976) Chicken embryo NAD levels lowered by teratogenic organophophorus and methylcarbamate insecticides. Biochem. Pharmacol., 25: 757-762.

QUARLES, J.M. & TENNANT, R.W. (1975) Effects of nitrosocarbaryl on BALB/3T3 cells. Cancer Res., 35: 2637-2645.

QUINBY, G.E. & BABIONE, R.W. (1966) Toxicological observations on the use of Baygon (2-isopropoxyphenyl N-methylcarbamate) as a malarial control insecticide in El Salvador. Ind. Med. Surg., 35: 596-597.

QURAISHI, M.S. (1972) Edaphic and water relationships of aldicarb and its metabolites. Can. Entomol., 104(3): 1191-1196.

REGAN, J.D., SETLOW, R.B., FRANCIS, A.A., & LYINSKY, W. (1976) Nitrosocarbaryl: its effect on human DNA. Mutat. Res., 38: 293-301.

REICH, G.A. & WELKE, J.O. (1966) Death due to a pesticide. New Engl. J. Med., 274: 1432.

REINER, E. (1971) Spontaneous reactivation of phosphorylated and carbamylated cholinesterases. Bull. World Health Org., 44: 109-112.

REINER, E. & ALDRIDGE, W.N. (1967) Effect of pH on inhibition and spontaneous reactivation of acetylcholinesterase treated with esters of phosphoric acids and of carbamic acids. Biochem. J., 105: 171-179.

REINER, E. & SKRINJARIC-SPOLJAR, M. (1968) Hydrolysis of some monomethylcarbamates in human sera. Croat. Chem. Acta, 40: 87-90.

REINER, E., BUNTIĆ, A., TRDAK, M., & SIMEON, V. (1974) Effect of temperature on the activity of human blood cholinesterases. Arch. Toxicol., 32: 347-350.

RHODES, R.C. & LONG, J.D. (1974) Run-off and mobility studies on benomyl in soils and turf. Bull. environ. Contam. Toxicol., 12(4): 385-393.

RICHARDSON, E.M. & BATTEESE, R.I., Jr (1973) An incident of Zectran poisoning. J. Maine Med. Assoc., 64: 158-159.

ROBEL, R.J., FELTHOUSEN, R.W., & DAYTON, A.D. (1982) Effects of carbamates on Bobwhite food intake body weight and locomotor activity. Arch. environ. Contam. Toxicol., 11: 611-615.

ROBENS, J.F. (1979) Teratologic Studies of Carbaryl, Diazinon, Norea, Disulfiram, and Thiram in small laboratory animals. Tox. Appl. Pharmacol., 15: 152-163.

ROMINE, R.R. (1973) Aldicarb. In: Sherma, J. & Zweig, G., ed. Analytical methods for pesticides and plant growth regulators, New York, London, Academic Press, Vol. 7, pp. 147-162.

ROSZKOWSKI, P. (1982) [Effect of the preparation propoxur on intrauterine development and on the state of bone formation in the progeny of the rat.] Ginek. Pol., 53(4): 201-207 (in Polish).

ROTHSCHILD, E.R., MANSER, R.J., & ANDERSON, M.P. (1982) Investigation of aldicarb in groundwater in selected areas of the central sand plain of Wisconsin. Groundwater, 20(4): 437-445.

RUZICSKA, P., PETER, S., & CZEIZEL, A. (1975) Studies on the chromosomal mutagenic effect of benomyl in rat and humans. Mutat. Res., 29: 201.

RYBAKOVA, M.N. (1967) [Comparative toxic effects of sevin and DDT used in the treatment of food crops.] Vopr. Pitan., 26: 9-15 (in Russian).

RYBAKOVA, M.N. (1968) [Effect of some pesticides on the hypophysis and its gonadotropic functions.] Gig. i Sanit., 33(11): 27-31 (in Russian).

SAKAI, K. & MATSUMURA, F. (1968) Esterases of mouse brain active in hydrolysing organophosphate and carbamate insecticides. J. agric. food Chem., 16(5): 803-807.

SAKAI, K. & MATSUMURA, F. (1971) Degradation of certain organophosphate and carbamate insecticides by human brain esterases. Toxicol. appl. Pharmacol., 19(4): 660-666.

SEILER, J.P. (1972) The mutagenicity of benzimidazole and benzimidazole derivatives. I. Forward and reverse mutations in Salmonella typhimurium caused by benzimidazole and some of its derivatives. Mutat. Res., 15: 273-276.

SEILER, J.P. (1975) Toxicology and genetic effects of benzimidazole compounds. Mutat. Res., 32: 151-168.

SEILER, J.P. (1976a) The mutagenicity of benzimidazole and benzimidazole derivatives. VI. Cytogenetic effects of benzimidazole derivatives in the bone marrow of the mouse and the Chinese hamster. Mutat. Res., 40: 339-347.

SEILER, J.P. (1976b) Inhibition of testicular DNA synthesis by chemical mutagens and carcinogens. Preliminary result in the validation of a novel short-term test. Mutat. Res., 46: 305-310.

SHAFIK, M.T., SULLIVAN, H.C., & ENOS, H.F. (1971) A method for the determination of 1-naphthol in urine. Bull. environ. Contam. Toxicol., 6: 34-39.

SHERMAN, H., CULIK, R., & JACKSON, R.A. (1975) Reproduction, teratogenic and mutagenic studies with benomyl. Toxicol. appl. Pharmacol., 32: 305-315.

SHTENBERG, A.I. & OZHOVAN, M.V. (1971) [Influence of low Sevin doses on the reproductive function of animals in a number of generations.] Vopr. Pitan., 30(1): 42-49 (in Russian, with English summary).

SHTENBERG, A.I. & RYBAKOVA, M.N. (1968) Effect of carbaryl on the neuroendocrine system of rats. Food Cosmet. Toxicol., 6: 461-467.

SHTENBERG, A.I., ORLOVA, N.V., & TORCHINSKY, A.M. (1973) [The action of different pesticides of different chemical structure on the gonads and embryogenesis of experimental animals.] Gig. i Sanit., 38(8): 16-20 (in Russian).

SIDEROFF, S.T. & SANTOLUCITO, J.A. (1972) Behavioural and physiological effects of cholinesterase inhibitor carbaryl (1-naphthyl methylcarbamate). Physiol. Behav., 9: 459-462.

SMALLEY, H.E., CURTIS, J.M., & EARL, F.L. (1968) Teratogenic action of carbaryl in beagle dogs. Toxicol. appl. Pharmacol., 13: 392-403.

SMALLEY, H.E., O'HARA, P.J., BRIDGES, C.H., & RADELEFF, R.D. (1969) The effects of chronic carbaryl administration on the neuromuscular system of swine. Toxicol. appl. Pharmacol., 14: 409-419.

SONAWANE, B.R. & KNOWLES, C.O. (1971) Phenmedipham and m-aminophenol decomposition in alkaline soil. Bull. environ. Contam. Toxicol., 6: 322-327.

SONAWANE, B.R. & KNOWLES, C.O. (1972) Comparative metabolism of two carbanilate herbicides (EP-475 and phenmedipham) in rats. Pestic. Biochem. Physiol., 1: 472-482.

STADNYK, L., CAMPELL, R.S., & JOHNSON, B.T. (1971) Pesticide effect on growth and ^{14}C assimilation in a fresh-water alga. Bull. environ. Contam. Toxicol., 6: 1-8.

STANSBURY, H.A., Jr & MISKUS, R. (1964) In: Zweig, G., ed. Analytical methods for pesticides, plant growth regulators, and food additives, New York, London, Academic Press, Vol. 2, pp. 437-450.

STEGEMAN LE ROY, C. (1964) The effects of the carbamate insecticide carbaryl upon forest soil mites and Collembola. J. econ. Entomol., 57(6): 803-808.

STRINGER, A. & LYONS, C.H. (1974) The effects of benomyl and thiophanate-methyl on earth worm populations; Apple Orchards Pestic. Sci., 5: 189-195.

STILL, G.G. & HERRETT, R.A. (1975) Methylcarbamates, carbanilates, and acylanilides. In: Kearney, P.C. & Kaufman, D.D., ed. Herbicides, chemistry, degradation, and mode of action. New York, Marcel Dekker, Vol. 2, pp. 609-664.

STILL, G.G. & MANSAGER, E.R. (1972) Metabolism of 4-chloro-2-butynyl-3-chloro-carbanilate by soybean plants. J. agric. food Chem., 20: 402-406.

STILL, G.G. & MANSAGER, E.R. (1973a) Soybean shoot metabolism of isopropyl 3-chlorocarbanilate: ortho and para arylhydroxylation. Pestic. Biochem. Physiol., 3: 87-95.

STILL, G.G. & MANSAGER, E.R. (1973b) Metabolism of isopropyl carbanilate by soybean plants. Pestic. Biochem. Physiol., 3: 289-299.

STILL, G.G. & MANSAGER, E.R. (1975) Alfalfa metabolism of propham. Pestic. Biochem. Physiol., 5: 515-522.

STILL, G.G. & RUSNESS, D.G. (1977) S-cysteinyl-hydroxychlorpropham: formation of the S-cysteinyl conjugate of isopropyl-3'-chloro-4'-hydroxy-carbanilate in oat (Avena savita L). Pestic. Biochem. Physiol., 7: 210-219.

STRINGER, A. & WRIGHT, M.A. (1973) The effect of benomyl and some related compounds on Lumbricus terrestris and other earthworms. Pestic. Sci., 4: 165-170.

STYLES, J.A. & GARNER, R. (1974) Benzimidazole carbamate methylester: evaluation of its effect in vivo and in vitro. Mutat. Res., 26: 177-187.

SULLIVAN, L.J., ELDRIDGE, J.M., KNAAK, J.B., & TALLANT, M.J. (1972) 5,6-dihydro-5,6-dihydroxycarbaryl glucuronide as a significant metabolite of carbaryl in the rat. J. agric. food Chem., 20(5): 980-985.

SUZUKI, T. & TAKEDA, M. (1976) Microbial metabolism of N-methylcarbamate insecticide I. Metabolism of O-sec-butylphenyl N-methylcarbamate by Aspergillus niger van Tieghem. Chem. Pharm. Bull., 24(9): 1967-1975.

SYPESTEYN, A.K., DEKHUYZEN, H.M., & VONK, J.W. (1977) Biological conversion of fungicides in plants and microorganisms, In: Siegel, M.R. & Sisler, H.D., ed. Antifungal compounds, New York, Marcel Dekker, Vol. 2, pp. 91-147.

TANNENBAUM, S.R., INSKEY, A.J., WEISMAN, M., & BISHOP, W. (1974) Nitrite in human. Its possible relationship to nitrosamine formation. J. Natl Cancer Inst., 53: 79-84.

TAYLOR, P. (1980) Anticholinesterase agents. In: Goodman, L.S. & Gilman, A., ed. The pharmacological basis of therapeutics, 6th ed., New York, Macmillan Publishing Co., pp. 100-119.

TOBIN, J.S. (1970) Carbofuran a new carbamate insecticide. J. occup. Med., 12: 16-19.

TOMLIN, A.D. & GORE, F.L. (1974) Effects of six insecticides and a fungicide on the numbers of biomass of earthworms in pasture. Bull. environ. Contam. Toxicol., 12(4): 487-492.

TORCHINSKY, A.M. (1973) [Significance of diets with different protein content in manifestations of teratogenic and embryotoxic action of some pesticides.] Vopr. Pitan., 3: 76-80 (in Russian).

TORCHINSKY, A.M., ORLOVA, N.V., SMIRNOVA, E.V., & MAKAROVA, L.F. (1976) [Data for toxico-hygienic characteristics of the pesticide Benlate of carbamate group.] Gig. i Sanit., 2: 35-39 (in Russian, with English summary).

UCHIYAMA, M., TAKEDA, M., SUZUKI, T., & YOSHIKAWA, K. (1975) Mutagenicity of nitroso derivatives of N-methylcarbamate insecticides in microbiological method. Bull. environ. Contam. Toxicol., 14: 389-394.

UNGERER, O., EISENBRAND, G., & PREUSSMANN, R. (1974) [The reaction of nitrite with pesticides.] Z. Krebsforsch., 81: 217-224 (in German, with English summary).

VANDEKAR, M. (1965) Observations on toxicity of carbaryl, folithion, and 3-isopropylphenyl-N-methylcarbamate in a village-scale trial in Southern Nigeria. Bull. World Health Org., 33: 107-115.

VANDEKAR, M., HEDAYAT, S., PLESTINA, R., & AHMADY, G. (1968) A study of the safety of O-isopropoxyphenyl methylcarbamate in an operational field-trial in Iran. Bull. World Health Org., 38: 609-623.

VANDEKAR, M., PLESTINA, R., & WILHELM, K. (1971) Toxicity of carbamates for mammals. Bull. World Health Org., 44: 241-249.

VAN ESCH, G.J. & KROES, R. (1972) Long-term toxicity studies of chlorpropham and propham in mice and hamsters. Food Cosmet. Toxicol., 10: 373-381.

VAN ESCH, G.J., VAN GENDEREN, H., & VINK, H.H. (1958) The production of skin tumours in mice by oral treatment with urethane, isopropyl-N-phenylcarbamate, or isopropyl-N-chlorophenyl carbamate in combination with skin painting with croton oil and Tween 60. Br. J. Cancer, 12: 355-362.

VASHAKIDZE, V.I. (1965) [Some questions of the harmful action of sevin on the reproductive function of experimental animals.] Soobshch. Akad. Nauk Gruz. SSR, 39: 471-478 (in Russian).

VASHAKIDZE, V.I. (1967) [Mechanism of action of pesticides (granosana, sevin, donoc) on the reproductive cycle of experimental animal.] Soobshch. Akad. Nauk Gruz. SSR, 48: 219-224 (in Russian).

VASSILIEFF, I. & ECOBICHON, D.J. (1982) Stability of Matacil® in aqueous media as measured by changes in anticholinesterase potency. Bull. environ. Contam. Toxicol., 29: 366-370.

VASSILIEFF, I. & ECOBICHON, D.J. (1983) Acute toxicity of aminocarb in male rats and inhibition of tissue esterases. Bull. environ. Contam. Toxicol., 31: 326-330.

VETTORAZZI, G. (1979) International regulatory aspects for pesticide chemicals. Vol. 1. Toxicity profiles, Boca Raton, Florida, CRC Press, Inc., 232 p..

VETTORAZZI, G. & VAN DEN HURK, G.W. (1984) Pesticide reference index, JMPR 1961-84, Geneva, World Health Organization (Unpublished report).

VOSS, G. & SCHULER, J. (1967) A fast and simple procedure for routine determination of plasma ChE activity. Bull. environ. Contam. Toxicol., 2(6): 357-363.

WHO (1984) The WHO recommended classification of pesticides by hazard. Guidelines to classification 1984-85, Geneva, World Health Organization (Unpublished report VBC/84.2).

WHO/FAO (1984) Data sheets on pesticides, Geneva, World Health Organization.

WEIL, C.S., WOODSIDE, M.D., CARPENTER, C.P., & SMYTH, H.F., Jr (1972) Current studies of tests of carbaryl for reproductive and teratogenic effects. Toxicol. appl. Pharmacol., 21: 390-404.

WEIL, C.S., WOODSIDE, M.D., BERNARD, J.B., CONDRA, N.I., KING, J.M., & CARPENTER, C.P., (1973) Comparative effect of carbaryl on rat reproduction and guinea-pig teratology when fed either in the diet or by stomach intubation. Toxicol. appl. Pharmacol., 26: 621-638.

WEIS, P. & WEIS, J.S. (1974) Schooling behaviour of Menidia menidia in the presence of insecticide Sevin (carbaryl). Mar. Biol., 28: 261-263.

WESTLAKE, G.E., BUNYAN, P.J., MARTIN, A.D., STANLEY, P.I., & STEED, L.C. (1981) Carbamate poisoning. Effect of selected carbamate pesticides on plasma enzymes and brain esterases of Japanese quail (Coturnix coturnix japonica). J. agric. food Chem., 29(4): 779-785.

WHITE, E.R., BOSE, E.A., OGAWA, J.M., MANJI, B.T., & KILGORA, W.W. (1973) Thermal and base-catalysed hydrolysis products of the systemic fungicide benomyl. J. agric. food Chem., 21: 616-618.

WIEDMANN, J.L.M., ECKE, G.G., & STILL, G.G. (1976) Synthesis and isolation of 1-hydroxy-2-propyl-3-chloro-carbanilate from soybean plants treated with isopropyl-3-chlorocarbanilate J. agric. food Chem., 24: 588-592.

WILHELM, K. & REINER, E. (1973) Effect of sample storage on human blood cholinesterase activity after inhibition by carbamates. Bull. World Health Org., 48: 235-238.

WILHELM, K., VANDEKAR, M., & REINER, E. (1973) Comparison of methods of measuring cholinesterase inhibition by carbamates. Bull. World Health Org., 48: 41-44.

WILLIAMS, I.H., BROWN, M.J., & WHITEHEAD, P. (1976a) Persistence of carbofuran residues in some British Columbia soils. Bull. environ. Contam. Toxicol., 15(2): 242-243.

WILLIAMS, I.H., PEPIN, H.S., & BROWN, M.J. (1976b) Degradation of carbofuran by soil microorganisms. Bull. environ. Contam. Toxicol., 15(2): 244-249.

WILLS, J.H., JAMESON, E. & COULSTON, F. (1968) Effects of oral doses of carbaryl on man. Clin. Toxicol., 1(3): 265-271.

WORTHING, C.R. & WALKER, S.B. (1983) The pesticide manual: a world compendium, 7th ed., London, British Crop Protection Council.

WRIGHT, M.A. & STRINGER, A. (1973) The toxicity of thiabendazole, benomyl, methyl-benzimidazol-2-yl-carbamate, and thiophanate-methyl to the earthworm Lumbricus terrestris. Pestic. Sci., 4: 431-432.

WYROBEK, A.J., WATCHMAKER, G., GORDON, L., WONG, K., MOORE, D.I., & WHORTON (1981) Sperm shape abnormalities in carbaryl exposed employees. Environ. Health Perspect., 40: 255-265.

YAKIM, V.S. (1967) [Data for substantiating the maximum permissible concentration of sevine in the air of working zones.] Gig. i Sanit., 32: 29-33 (in Russian).

YOSHIDA, K. & NISHIUCHI, Y. (1972) [Toxicity of pesticides to some water organisms.] Bull. agric. Chem. Inspect. Stn. (Japan), 12: 122-128 (in Japanese).

YOSHIDA, K. & NISHIUCHI, Y. (1976) [Toxicity of pesticides to some aquatic animals.] Bull. agric. Chem. Inspect. Stn. (Japan),. 6: 65-69 (in Japanese).

ZAKI, M.H., MORAN, D., & HARRIS, D. (1982) Pesticides in groundwater: the aldicarb story in Suffolk Country, New York. Am. J. public Health, 72(12): 1391-1395.

Annex I. Names and structures and some physical and chemical properties of carbamate pesticides

Common name	Trade/other name	Chemical structure	CAS chemical name/ CAS registry number	Molecular formula	Relative molecular mass	Vapour pressure (25 °C)	Water solubility (25 °C)
aldicarb	Temik	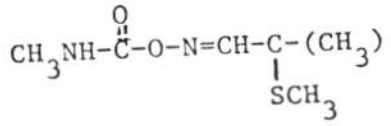	propanal, 2-methyl-2-(methylthio)-, O-[(methylamino) carbonyl]oxime (116-06-3)	$C_7H_{14}N_2O_2S$	190.29	13 mPa	0.6%
aldoxycarb	Standak	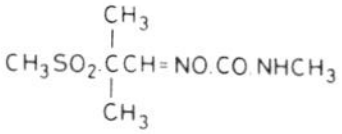	2-methyl-2-(methylsulfonyl)-propanal O-[(methylamino)-carbonyl]oxime (1646-88-4)	$C_7H_{14}N_2O_4S$	222.3	12 mPa	0.9%
allyxycarb	Hydrol APC	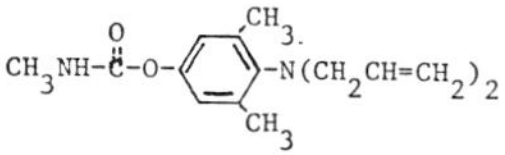	phenol, 4-(di-2-propenylamino)-3,5-dimethyl, methyl carbamate (6392-46-7)	$C_{16}H_{22}N_2O_2$	274.4		
aminocarb	Matacil	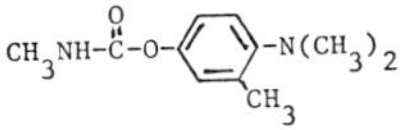	phenol, 4-(dimethylamino)-3-methyl-, methyl carbamate (2032-59-9)	$C_{11}H_{16}N_2O_2$	208.29	non-volatile	slightly
asulam	Asulox Asilan	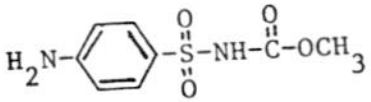	carbamic acid, [(4,-aminophenyl)-sulfonyl], methyl ester (3337-71-1)	$C_8H_{10}N_2O_4S$	230.26	-	0.5%

Annex I (contd).

Common name	Trade/other name	Chemical structure	CAS chemical name/ CAS registry number	Molecular formula	Relative molecular mass	Vapour pressure (25 °C)	Water solubility (25 °C)
barban	CBN Carbyne Chlorinat	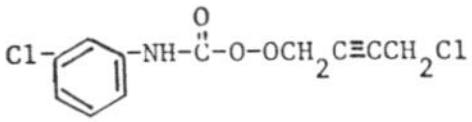	carbamic acid, 3-chlorophenyl-, 4-chloro-2-butynyl ester (101-27-9)	$C_{11}H_9Cl_2NO_2$	258.11	50 µPa	0.0011%
BPMC	Bay-Bassa Baycarb Sumibas	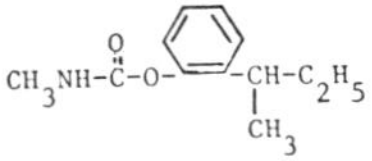	phenol, 2-(1-methyl-propyl)-,methyl carbamate (3766-81-2)	$C_{12}H_{17}NO_2$	207.3	48 mPa[b]	0.061%[c]
bendiocarb	Ficam Garvox Multimet Tattoo Seedox	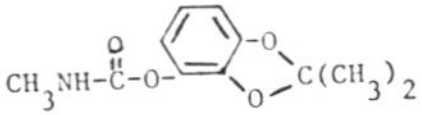	1,3-benzodioxol-4-yl, 2,2-dimethyl, methyl carbamate (22781-23-3)	$C_{11}H_{13}NO_4$	223.25	660 µPa	0.004%
benomyl	Arilate Benlate Tersan	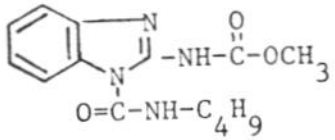	carbamic acid,[1-[(butylamino)-carbonyl]-1H-benzimidazole-2-yl]-, methyl ester (17804-35-2)	$C_{14}H_{18}N_4O_3$	290.36	non-volatile	~0.0002%
bufencarb	Bux Metalkamate	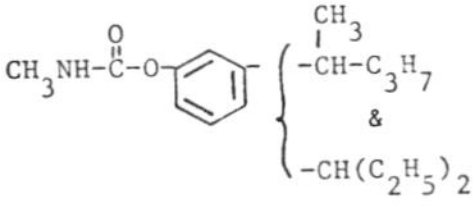	mixture of phenol, 3-(1-methylbutyl)-phenyl)-, and 3-(1-ethylpropyl-phenyl)-, methyl carbamates (8065-36-9)	$C_{13}H_{19}NO_2$	221.33	4.0 mPa[c]	< 0.005%

Annex I (contd).

butacarb		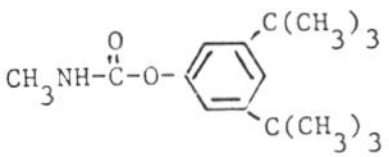	phenol, 3,5-bis (1,1-dimethylethyl)-, methyl carbamate (2655-19-8)	$C_{16}H_{25}NO_2$	263		
carbanolate	Banol SOK	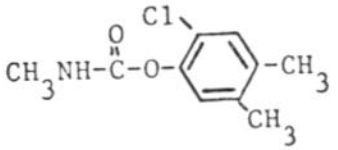	phenol, 2-chloro-4, 5-dimethyl-, methyl carbamate (671-04-5)	$C_{10}H_{12}ClNO_2$	213.68		
carbaryl	Sevin (Sevkul) (Carbicide)	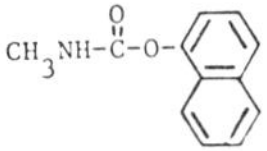	1-naphthalenol methyl carbamate (63-25-2)	$C_{12}H_{11}NO_2$	201.24	< 665 mPa[f]	0.012%[c]
carbendazim	Bavistin BCM/MBC Derosal	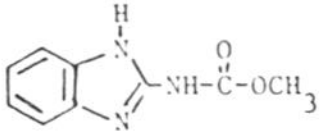	carbamic acid, -1H-benzimidazole-2-yl-, methyl ester (10605-21-7)	$C_9H_9N_3O_2$	191.18	< 100 nPa[b]	0.0028%[c] (pH 4.8)
carbetamide	Legurame Carbetamex	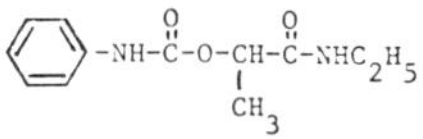	propanamide, N-ethyl -2-[[(phenyl-amino) carbonyl]oxy]-, (R)- (16118-49-3)	$C_{12}H_{16}N_2O_3$	236.30	negligible at 20°C	0.35% [b]
carbofuran	Furadan Yaltox Curater	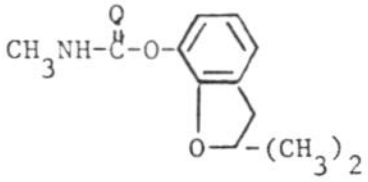	7-benzofuranol, 2,3-dihydro-2,2-dimethyl-, methyl carbamate (1563-66-2)	$C_{12}H_{15}NO_3$	221.28	2.7 mPa[g]	0.07%
chlorbufam	-	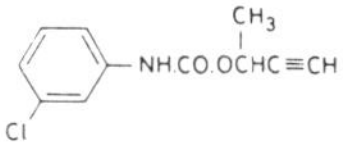	1-methylprop-2ynyl-3-chlorocarbanilate	$C_{11}H_{10}ClNO_2$	223.7	2.1 mPa[b]	0.054%[b]

Annex I (contd).

Common name	Trade/other name	Chemical structure	CAS chemical name/ CAS registry number	Molecular formula	Relative molecular mass	Vapour pressure (25 °C)	Water solubility (25 °C)
cloethocarb	-	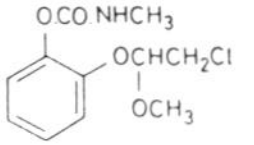	2-(2-chloro-1-[a] methoxyethoxy)- phenylmethyl carbamate	$C_{11}H_{14}ClNO_4$	259.7	-	0.13%[b]
chlorpropham	Elbanil Furloe Prevenol Chloro IPC (CIPC)	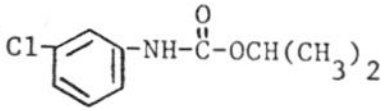	carbamic acid, 3-chlorophenyl-, 1-methylethyl ester (101-21-3)	$C_{10}H_{12}ClNO_2$	213.68	-	0.009%
desmedipham	Betanex	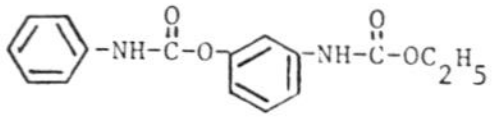	carbamic acid, [3-[[(phenylamino)-carbonyl]oxy]phenyl]-, ethyl ester (13684-56-6)	$C_{16}H_{16}N_2O_4$	300.34		
dimetilan	Snip	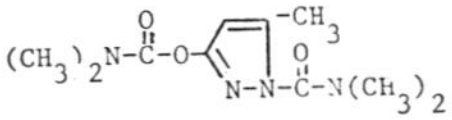	carbamic acid, dimethyl-, 1-[(dimethylamino) carbonyl]-5-methyl-1H-pyrazol-3-yl ester (644-64-4)	$C_{10}H_{16}N_4O_3$	240.3		readily
dichlormate		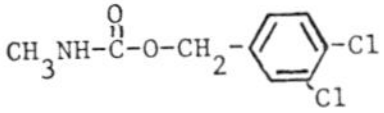	benzenemethanol, 2,3-dichloro-, methyl carbamate (62046-37-1)	$C_9H_9Cl_2NO_2$	199		0.017%

Annex I (contd).

dioxacarb	Famid Elecron	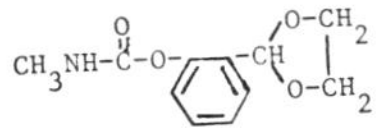	phenol, 2-(1,3-di-oxolan-2-yl)-, methyl carbamate (6988-21-2)	$C_{11}H_{13}NO_4$	223.25	40 µPa[c]	0.6%[c]
ethiofencarb	Croneton	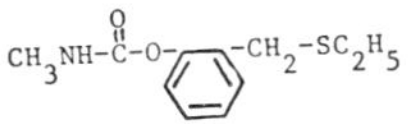	phenol, 2-[(ethyl-thio)methyl]-, methyl carbamate (29973-13-5)	$C_{11}H_{15}NO_2S$	225.33	13 mPa[c]	0.182%[b]
etrofol	hopcide CPMC	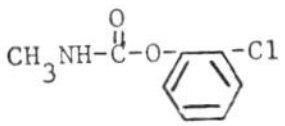	phenol, 2-chloro-, methyl carbamate (3942-54-9)	$C_8H_8ClNO_2$	185.62	-	
formetanate	Dicarzol Carzol	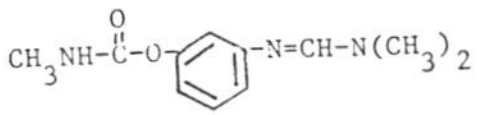	methanimidamide, N,N-dimethyl-N'-[3-[[(methylamino) carbonyl]oxy]-phenyl]- (22259-30-9)	$C_{11}H_{15}N_3O_2$	257.75	27 µPa	0.63%
isoprocarb	Etrofolan Carbamat IPM Hytox Mipcin MIPC	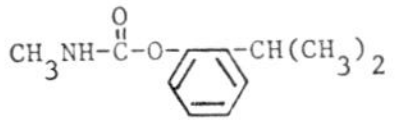	phenol, 2-(1-methyl-ethyl)methyl carbamate (2631-40-5)	$C_{11}H_{15}NO_2$	193.27	-	insoluble
karbutilate	Tandex	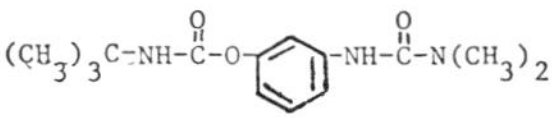	carbamic acid, (1,1-dimethylethyl)-, 3-[[(dimethylamino) carbonyl]-amino] phenyl ester (4849-23-5)	$C_{14}H_{21}N_3O_3$	279.38	6.0 nPa[b]	0.033%[b]

Annex I (contd).

Common name	Trade/other name	Chemical structure	CAS chemical name/ CAS registry number	Molecular formula	Relative molecular mass	Vapour pressure (25 °C)	Water solubility (25 °C)
MPMC	Meobal	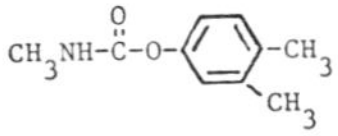	phenol, 3,4-dimethyl-, methyl carbamate (2425-10-7)	$C_{10}H_{13}NO_2$	179.24		
methiocarb	Draza Mesurol	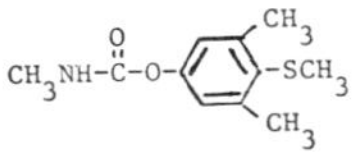	phenol, 3,5-dimethyl-4-(methylthio)-, methyl carbamate (2032-65-7)	$C_{11}H_{15}NO_2S$	225.33	15 mPa[h]	0.003%[b]
methomyl	Lannate Methavin	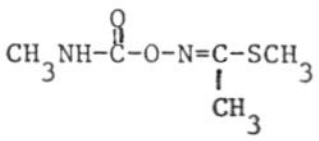	ethanimidothioic acid, N-[[(methylamino)carbonyl]oxy]-, methyl ester (16752-77-5)	$C_5H_{10}N_2O_2S$	162.23		5.8%
metolcarb	-	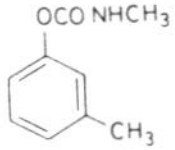	m-tolyl methyl[a] carbamate	$C_9H_{11}NO_2$	165.2	145 mPa[b]	0.26%[c]
mexacarbate	Zectran	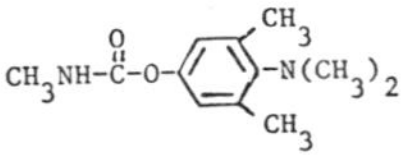	phenol, 4-dimethylamino)-3,5-dimethyl-, methyl carbamate (ester) (315-18-4)	$C_{12}H_{18}N_2O_2$	222.32		

Annex I (contd).

oxamyl	Vydate	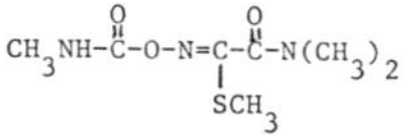	ethanimidothioic acid, 2-(dimethyl-amino)-N-[[(methyl-amino)carbonyl]-oxy]-2-oxo-, methyl ester (23135-22-0)	$C_7H_{13}N_3O_3S$	219.29	31 mPa	28.0%
phenmedipham	Betanal	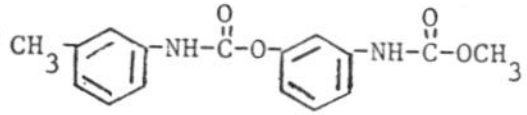	carbamic acid, (3-methylphenyl)-, 3-[(methoxycarbonyl) amino]phenyl ester (13684-63-4)	$C_{16}H_{16}N_2O_4$	300.34	-	~0.0005%
pirimicarb	Aphox Fernos Pirimor Rapid	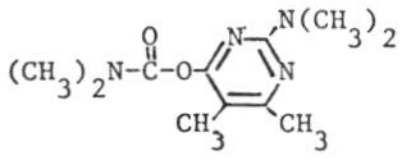	carbamic acid, dimethyl-, 2-(di-methylamino)-5, 6-dimethyl-4-pyrimi-dinyl ester (23103-98-2)	$C_{11}H_{18}N_4O_2$	238.33	4.0 mPa[c]	0.27%
promacyl	Promicide	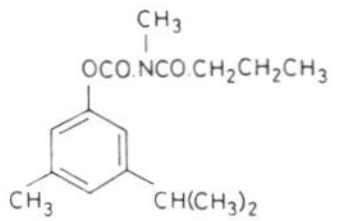	5-methyl-m-cumenyl[a] butyryl(methyl) carbamate	$C_{16}H_{23}NO_3$	277.4	400 Pa[i]	slightly
promecarb	Carbamult Minacide	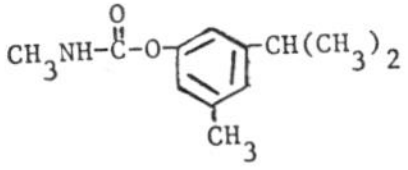	phenol, 3-methyl-5-(1-methylethyl)-, methyl carbamate (2631-37-0)	$C_{12}H_{17}NO_2$	207.3		0.0092%
propham	Banhoe Tuberit IPC	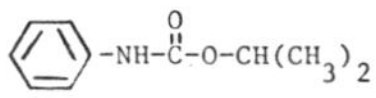	carbamic acid, phenyl-, 1-methyl-ethyl ester (122-42-9)	$C_{10}H_{13}NO_2$	179.24	-	0.025%

Annex I (contd).

Common name	Trade/other name	Chemical structure	CAS chemical name/ CAS registry number	Molecular formula	Relative molecular mass	Vapour pressure (25 °C)	Water solubility (25 °C)
propoxur	Baygon Blattanex Sendran Suncide Unden Tendex Arprocarb	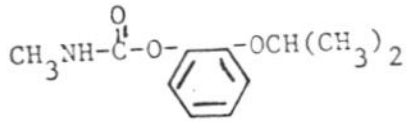	phenol, 2-(1-methyl ethoxy)-, methyl carbamate (114-26-1)	$C_{11}H_{15}NO_3$	209.27	6.5 x 10^{-6} mm Hg (20 °C)	0.2%
swep		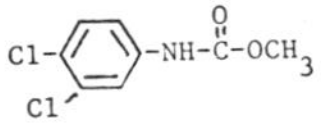	carbamic acid, 3,4-dichlorophenyl-, methyl ester (1918-18-9)	$C_8H_7Cl_2NO_2$	220.06		
terbucarb	Azac Azar		2,6-di-ter-butyl-p-tolyl methyl carbamate	$C_{17}H_{27}NO_2$	277.4		0.0067%
thiodicarb	Larvin		ethanimidothioic acid, N,N'-[thiobis (methylimino)-carbonyloxy] bis-dimethyl-ester (59669-26-0)	$C_{10}H_{18}N_4O_4S_3$	354.3		
thiofanox	Dacamox	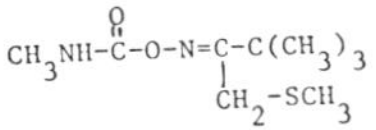	2-butanone, 3,3-dimethyl-1-(methylamino)-, O-[(methylamino)- carbonyl]oxime (39196-18-4)	$C_9H_{18}N_2O_2S$	218.35	22.6 mPa	0.52%[d]

Annex I (contd).

thiophanate ethyl	Topsin Cercobin Enovit	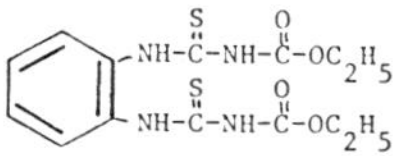	carbamic acid, [1,2-phenylenebis (iminocarbonothioyl)] bis-, diethyl ester (23564-06-9)	$C_{14}H_{18}N_4O_4S_2$	370.48	-	insoluble
thiophanate methyl	Topsin-M Cercobin-M Enovit-M Fungo Mildothane	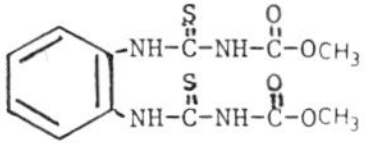	carbamic acid, [1,2-phenylenebis-iminocarbonothioyl)] bis-, dimethyl ester (23564-05-8)	$C_{12}H_{14}N_4O_4S_2$	342.42	-	slightly
trimethacarb	Broot Landrin		4:1 mixture of phenol, 3,4,5-tri-methyl-, methyl-carbamate and phenol, 2,3,5-trimethyl-methyl carbamate (2686-99-9 and 2655-15-4)	$C_{11}H_{15}NO_2$	193.24		
MTMC	Tsumacide	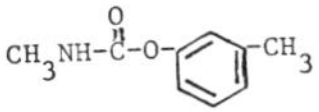	carbamic acid, methyl-, 3-methyl-phenyl ester (1129-41-5)	$C_9H_{11}NO_2$	165.21		
xylylcarb		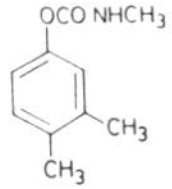	3,4-xylylmethyl[a] carbamate	$C_{10}H_{13}NO_2$	179.2	70 mPa[b]	0.058%[b]
XMC	Macbal Cosban		3,5 xylylmethyl[a] carbamate	$C_{10}H_{13}NO_2$	179.2		insoluble

a IUPAC name.
b At 20 °C.
c At 30 °C.
d At 22 °C.
e From: Worthing & Walker (1983) (given in the reference list of the main document).
f At 26 °C.
g At 33 °C.
h At 60 °C.
i At 149 °C.

Annex II. Summary of the short- and long-term toxicity studies that were used to establish the acceptable daily intakes for human beings for carbamate compounds

Carbamate	Duration of test	No-observed-adverse-effect level in rats[a]		Reference
		mg/kg diet	mg/kg body weight	
Aldicarb	2 years	2.5	0.125	FAO/WHO (1983b)
			0.25 (dog)	
Aminocarb	2 years	100		FAO/WHO (1980b)
Bendiocarb			0.38	
			0.25 (dog)	FAO/WHO (1985a)
Benomyl	2 years	2500	125	FAO/WHO (1983b, 1985a)
			30 (based on teratology)	
Carbaryl	2 years	200	10	FAO/WHO (1985a)
	6 weeks		0.06 (man)	
	80 weeks	400 (mice)	60 (mice)	
	1 year		1.8 (dog)	
Carbendazim	2 years	500	25	FAO/WHO (1986b)
	2 years	100 (dog)	2.5	
Carbofuran	2 years	20	1.0	FAO/WHO (1981b)
	2 years	20 (mice)	2.5	

Annex II (contd).

Chloropropham	2 years	2000		FAO/WHO (1965b)
Ethiofencarb	90 days	500 1000 (dog)	10 25	FAO/WHO (1983b)
Methiocarb	1 year	25 5 (dog)	1.3 0.125	FAO/WHO (1985a)
Methomyl	22 months	100	5	FAO/WHO (1978b)
Oxamyl	14 days	100 (dog)	2.5	FAO/WHO (1985a)
Pirimicarb	2 years 2 years 17 weeks	175	9 1.8 (dog) 2 (monkey)	FAO/WHO (1983b)
Propoxur	2 years 2 years	250	12.5 (dogs)	FAO/WHO (1983b)
Thiophanate-methyl	2 years 2 years 2 years	160 160	8 50 (dog) 23 (mice)	FAO/WHO (1976b, 1978b)

a Data given for rats, unless otherwise stated.
References are to reports of WHO/FAO Joint Meeting on Pesticide Residues. Vettorazzi (1979) gives a summary of these reports up to and including 1977.

Note: The references to Annex II are given in the reference list of the main document.

Annex III. Carbamates: JMPR reviews, ADIs, Evaluation by IARC, Classification by Hazard, FAO/WHO Data Sheets, IRPTC Data Profile and Legal File[a]

Compound	Year of JMPR meeting	ADI[b] (mg/kg body weight)	Evaluation by JMPR[c]: Published in: FAO/WHO	IARC[d] Evaluation of Carcino-genicity	Availability of IRPTC[e]: Data Profile	Availability of IRPTC[e]: Legal file[f]	WHO recom-mended clas-sification of pesticides by hazard[g]	FAO/WHO Data Sheets on Pesticides[h]
Aldicarb	1982	0-0.005	1983b		+	+	IA	No. 53 (1982)
			1983a					
	1979	0-0.001	1980b					
			1980a					
Aminocarb	1979	no ADI	1980b				IB	
			1980a					
	1978	no ADI	1979b					
			1979a					
	1984	0-0.004	1985b					
Bendiocarb	1982	0-0.002 (temporary)	1983b		+	+	II	No. 52 (1982) (rev. 1)
Benomyl	1983	0.02	1984		-	+		No. 87 (in
			1985a				0	(preparation)
	1978	no ADI	1979a					
	1975	no ADI	1976a					
			1976a					
	1973	no ADI	1974a					
			1974b					
Carbaryl	1984[i]	0-0.01	1985b					
	1979[i]	0-0.01	1980b		+	+	II	No. 3 (1978) (rev. 1)
			1980a					
	1977[i]	0-0.01	1978b					
			1978a					
	1976[i]	0-0.01	1977b	Vol. 12, page 37				

Annex III (contd).

Carbaryl (contd)	1975[i]	0-0.01	1977a 1976b 1976a		
	1973	0-0.01	1974b 1974a		
	1971[i]	0-0.01 (temporary)	1972a		
	1970[i]	0-0.01 (temporary)	1971b		
	1969	0-0.01 (temporary)	1971a 1970b		
	1968[i]	0-0.02	1970a 1969b 1969a		
	1967	0-0.02	1968b 1968a		
	1966	0-0.02	1967b 1967a		
	1965	0-0.02	1965b 1965a		
	1963	0-0.02	1964		
Carbendazim	1985	0-0.01	1986b		
	1983	0-0.01	1984 1985a	0	No. 89 (in (preparation)
	1978	no ADI	1979a 1979b		
	1977	no ADI	1978a 1978b		
	1976	no ADI	1977a 1977b		
	1973	no ADI	1974a 1974b		

Annex III (contd).

Compound	Year of JMPR meeting	ADI[b] (mg/kg body weight)	Evaluation by JMPR[c]: Published in: FAO/WHO	IARC[d] Evaluation of Carcinogenicity	Availability of IRPTC[e]: Data Profile	Legal file[f]	WHO recommended classification of pesticides by hazard[g]	FAO/WHO Data Sheets on Pesticides[h]
Carbofuran	1982	0-0.01	1983a		+	+	IB	
	1980	0-0.01	1981b					
			1981a					
	1979	0-0.003 (temporary)	1980b					
			1980a					
	1976	0-0.003 (temporary)	1977b					
			1977a					
Chlorpropham	1965	no ADI	1965b	Vol. 12, page 55			0	
			1965a					
	1963	no ADI	1964					
Ethiofencarb	1983[i]	0-0.1	1984a				II	
	1982	0-0.1	1983b					
			1983a					
	1981[i]	0-0.1 (temporary)	1982b					
			1982a					
	1980[i]	0-0.1 (temporary)	1981b					
			1981a					

Annex III (contd).

Ethiofencarb (contd)	1978[i]	0-0.1 (temporary)	1979b					
			1979a					
	1977	0-0.1 (temporary)	1978b					
Methiocarb	1985	0-0.001	1986b					
	1984	0-0.001	1985b				II	
	1983	0.001	1984					
			1985a					
	1981	0-0.001	1982b					
			1982a					
Methomyl	1978	no ADI	1979b		+	+	IB	No. 55 (1983)
			1979a					
	1977[i]	no ADI	1978b					
			1978a					
	1976[i]	no ADI	1977b					
			1977a					
	1975[i]	no ADI	1976b					
			1976a					
Mexacarbate				Vol. 12, page 237				
Oxamyl	1985	0-0.03	1986b					
	1984	0-0.03	1985b					
	1983[i]	0-0.01 (temporary)	1984		+	+	IB	No. 54 (1983)
	1980	0-0.01 (temporary)	1981b					
			1981a					
		0-0.0014						

Annex III (contd).

Compound	Year of JMPR meeting	ADI[b] (mg/kg body weight)	Evaluation by JMPR[c]: Published in: FAO/WHO	IARC[d] Evaluation of Carcinogenicity	Availability of IRPTC[e]: Data Profile	Legal file[f]	WHO recommended classification of pesticides by hazard[g]	FAO/WHO Data Sheets on Pesticides[h]
Pirimicarb	1982	0-0.02	1983b 1983a		+	+	II	
	1981[i]	0-0.01 (temporary)	1982b 1982a					
	1979[i]	0-0.01 (temporary)	1980b 1980a					
	1978	0-0.01 (temporary)	1979b 1979a					
	1976	0-0.004 (temporary)	1977b 1977a					
Propham	1965	no ADI	1965a 1965b	Vol. 12 Page 189	+	+	0	
	1963	no ADI	1964a					
Propoxur	1983	0-0.02	1984		+	+	II	No. 25 (1976)
	1981[i]	0-0.02	1982b 1982a					
	1978[i]	0-0.02	1979a					
	1977[i]	0-0.02	1978a					
	1973	0-0.02	1974b 1974a					
Thiodicarb	1985	0-0.1 (temporary)	1986b					

Annex III (contd).

Thiofanox	1978[i]	no ADI	1979a		IB
Thiophanate-methyl	1978[i]	0-0.08	1979a	+	0
	1977	0-0.08	1978b		
			1978a		
	1975	0-0.08	1976b		
			1976a		
	1973	0-0.08	1974b		
			1974a		

a Adapted from: Vettorazzi & van den Hurk (1984).
b ADI = acceptable daily intake.
c JMPR = Joint Meeting on Pesticide Residues (FAO/WHO).
d IARC = International Agency for Research on Cancer (WHO, Lyons, France).
e IRPTC = International Register of Potentially Toxic Chemicals (UNEP, Geneva).
f From: IRPTC (1983).
g From: WHO (1984a). See this reference for classification of organophosphates not mentioned in this annex.
The hazard referred to in this Classification is the acute risk for health (that is, the risk of single or multiple exposures over a relatively short period of time) that might be encountered accidentally by any person handling the product in accordance with the directions for handling by the manufacturer or in accordance with the rules laid down for storage and transportation by competent international bodies.
Classification relates to the technical material, and not to the formulated product:

Class		LD_{50} for the rat (mg/kg body weight) Oral		Dermal	
		Solids	Liquids	Solids	Liquids
1A	Extremely hazardous	5 or less	20 or less	10 or less	40 or less
1B	Highly hazardous	5 - 50	20 - 200	10 - 100	40 - 400
II	Moderately hazardous	50 - 500	200 - 2000	100 - 1000	400 - 4000
III	Slightly hazardous	over 500	over 2000	over 1000	Over 4000
0	Unlikely to present acute hazard in normal use				

h WHO/FAO Data Sheets on Pesticides with number and year of appearance.
i No toxicological evaluation - residues only.
N.B. References to Annex II are listed in the reference list of the main document.

Annex IV. Abbreviations

2-AB	2-aminobenzimidazole
ACh	acetylcholine
AChE	acetylcholinesterase
ACTH	adrenocorticotropic hormone
ADI	acceptable daily intake
ChE	cholinesterase
CNS	central nervous system
EEG	electroencephalogram
EMG	electromyogram
FAO	Food and Agriculture Organization of the United Nations
4-HBC	4-hydroxy-2-benzimidazole carbamate
5-HBC	5-hydroxy-2-benzimidazole carbamate
IARC	International Agency for Research on Cancer
im	intramuscular
IPCS	International Programme on Chemical Safety (World Health Organization)
IRPTC	International Register of Potentially Toxic Chemicals (United Nations Environment Programme)
iv	intravenous
JMPR	Joint Meeting of the FAO Panel of Experts on Pesticide Residues in Food and the Environment and a WHO Expert Group on Pesticide Residues
MFO	mixed-function oxidase
MLD	minimum lethal dose
MRL	maximum residue limit

NTE	neuropathy target esterase (formerly neurotoxic esterase)
PBI	protein-bound iodine
pseudoChE	pseudocholinesterase
RBC	red blood cells
sc	subcutaneous
UVR	ultraviolet radiation